E-Book inside

Please register on:
https://www.trackomedia.com/e-book-download
and use this publication as an e-book for free.

The download code for this book is: **MGBRDY**

ETCS
for Engineers

Ian Mitchell (Editor)

2nd edition

TrackoMedia

About the IRSE

The Institution of Railway Signal Engineering is the international professional body for all those engaged in or associated with railway signalling and telecommunications, train control, traffic management and allied professions. The institution was formed in 1912 with the aims of advancing the science and practice of train control and maintaining high standards of knowledge and competence within the profession. The headquarters is in London, UK and there are more than 20 local sections in all the continents of the world where railways operate. The IRSE organises a wide range of activities to share knowledge and support members' professional development via online and in-person lectures, conferences and technical visits, and by publishing textbooks and a monthly magazine. It also operates professional qualification, registration and licensing schemes that support railway administrations in their assurance of staff competence.

German National Library bibliographic information:

The German National Library (Deutsche Nationalbibliothek) catalogues this publication in the German National Bibliography (Deutsche Nationalbibliografie), detailed bibliographic information can be found on http://dnb.de

Publishing House:	GRT Global Rail Academy and Media GmbH Werkstättenstraße 18 51379 Leverkusen, Germany Phone: +49 (0) 40 – 228679-500 Web: www.trackomedia.com E-Mail: office@globalrailmedia.com
Managing Director:	Detlev K. Suchanek
Editorial Office:	Alexandra Schöner
Distribution/Book Service:	Sabine Braun
Advertisements:	Dirk Bogisch
Cover Design:	Shujun Wong, Hamburg
Cover Image:	ETCS Driver Machine Interface example from ERA_ERTMS_015560
Print:	TZ-Verlag & Print GmbH, Roßdorf

2nd edition 2026
ISBN 978-3-96245-271-1

A publication by

TrackoMedia

TrackoMedia is the publishing brand of
GRT Global Rail Academy and Media GmbH.

Preface

Professor Rod Muttram FREng, FIRSE

I am proud to be the 102nd President of the IRSE for the year 2026-2027.

The IRSE is the professional institution for all those engaged in, or associated with, railway signalling and telecommunications, train control, traffic management and the allied professions.

Our recent name change from the Institution of Railway Signal Engineers to the Institution of Railway Signal Engineering reflects the wide range of disciplines now involved in safely and efficiently controlling all forms of guided transportation.

The European Train Control System (ETCS) is typical of the current technology used in the railway industry, depending on many of those disciplines; computer hardware, software, communications, cyber security, operations, human factors, specialised components (and increasingly machine learning) together with deep domain knowledge of how they all fit together to deliver the unique requirements of controlling and regulating trains.

I have been involved with ETCS for many years, as Chairman of The European Rail Research Institute (ERRI) when the initial specifications were drafted there in the early 1990s, and later as a board member of the Association Européenne pour l'Interoperabilité Ferroviaire (AEIF) and of the ERTMS Users Group (the European Rail Traffic Management System of which ETCS is part). My signature is one of those on the original ERTMS/ETCS Class 1 Specifications delivered to the European Commission in April 2000. In the intervening 25 years, ETCS has evolved and matured into a highly reconfigurable train control system including additional features such as Automatic Train Operation (ATO).

Revising this textbook under Ian Mitchell's leadership to reflect those many changes has been both challenging and rewarding, and I hope it will prove helpful to those coming to ETCS for the first time or seeking a deeper understanding of this complex system.

Editor's Introduction

Ian Mitchell

The Institution of Railway Signal Engineering (IRSE) has a long tradition of publishing textbooks to support the professional development of railway engineers. As ETCS emerged as an important element of the railway signalling world it was obvious that a textbook covering the topic would be required, and the decision was taken to partner with an established European publisher of railway literature to fulfil the need. The outcome of this initiative was the first edition of this book 'ETCS for Engineers' published in 2010. The title was deliberately chosen to emphasise that the aim is to describe the technical detail of ETCS, and not the commercial and political aspects of its development and deployment on projects.

The editor for the first edition was the late and much respected former IRSE President Peter Stanley. He assembled a remarkable international team of authors, many of whom had been deeply involved in the early development of ETCS. The resulting book took the form of an extended series of essays describing all aspects of the ETCS system as it existed at that time.

When I was asked to lead the initiative to produce a second edition, my initial assumption was that this would simply be an update of the existing text to address the development of ETCS specifications in the intervening years. However, it soon became apparent that this would not be straightforward, as many of the original authors had retired or moved to different roles. There was also a certain element of overlap between their contributions. The decision was taken to undertake a more fundamental revision, and while some of the original text remains, other parts have been re-written from scratch.

High-level information about ETCS is readily available in magazine articles and websites, and the detailed specifications are also publicly available from the European Agency for Railways (ERA). Our aim with this book is to fill the gap between these, explaining the technical detail in a manner that will be understandable to an engineer who is new to ETCS. I hope very much that we will fulfil that need.

After much consideration, we decided to avoid describing specific projects that have deployed ETCS technology. To do justice to the wide range of applications that ETCS makes possible would double the size of the book. Fortunately, our publishers have a parallel project to fill this gap with another book, 'ETCS in Europe', which will describe applications in several European countries.

Acknowledgements

This new edition has been several years in the making, and I am very fortunate to have been supported throughout by a small working group of IRSE members. **Rod Muttram** and **Tom Godfrey** have written a substantial part of the new text, while **Frans Heijnen** and **Jacques Poré** provided continuity as they had contributed to the first edition.

I would also like to thank IRSE members **Francis How** for proof-reading and **Mark Glover** who produced many of the figures, and **Alexandra Schöner**, our professional editor at Global Rail Academy and Media who has taken the output from engineers and managed the process to make it into a real book.

Whilst this is in no way an official publication, we have had informal support from the ERA and UNISIG and have reproduced some figures from their published standards.

Finally, I would also like to acknowledge the following original authors of the first edition and others who provided us with input for the new version. I hope they are not too surprised by the level of change from their original contribution and take full responsibility for the end result.

Norbert Apel
Per Axelsson
Ron Bailes
Wim Coenraad
John Harmer
Carsten Hasselkuss
Georg Hemzal
Stephan Hinze
Jorge Iglesias
Hans-Georg Kast
Karin Löfstedt
Franciso Lozano
Per Lundberg
J. Mattisson
Olaf Mense
Michael Meyer zu Hörste
Fernado Montes
Marcus Montigel
Patrice Noury
Oskar Stalder
Bernhard Stamm
Jenny Strahl
Jaime Tamarit
Michael Watkins
Peter Woodbridge
Patrick Zoetardt

Part 1
Introduction to ETCS

The first part of the book provides a brief overview of ETCS, its relationship with wider concepts such as ERTMS and Technical Specifications for Interoperability, and the benefits to railway operations that these initiatives are intended to deliver. The history of ETCS development over more than 30 years is described, with an explanation of the wide range of stakeholders that have been involved in establishing it as a European and World standard. The progressive enhancement of the system to incorporate additional 'game changer' functionality is described, and the realisation of the anticipated benefits is reviewed.

Chapter 1: What is ETCS?

1.1 ETCS and ERTMS

The terms ETCS (European Train Control System) and ERTMS (European Rail Traffic Management System) are often used interchangeably, and the composite term ERTMS/ETCS is also frequently seen. They are, however, not the same:

ETCS is a highly configurable suite of mainline railway train control equipment built to a set of specifications managed by the European Union Agency for Railways (ERA). It is available from multiple suppliers, and its role is to:

- Communicate a movement authority from the trackside to the train
- Present movement authority information to the driver in the train cab
- Provide on-board automatic train protection (ATP) functionality to supervise the safe movement of the train

ERTMS is the global term for the railway traffic management systems that are being standardised by the European Union. These include ETCS, Railway Mobile Radio (RMR) and Automatic Train Operation (ATO). RMR is the generic term for systems that provide ETCS track-train data transmission and voice communications. It includes GSM-R and FRMCS (Future Railway Mobile Communication System), which is GSM-R's future replacement.

A key feature of all the ERTMS specifications is that they define the track to train interface and functionality in sufficient detail to ensure that equipment from different suppliers will seamlessly exchange and use information.

This textbook focuses on ETCS but also includes information on how ETCS interacts with the other components of ERTMS.

1.2 Introduction to ETCS levels

The standardised ETCS on-board equipment allows a train to operate over different levels of ETCS trackside implementation. Originally three main levels were defined, the key difference between them being how the trackside and on-board systems communicated with each other. Level 1 used entirely track-based communication, and levels 2 and 3 used radio communications. The difference between level 2 and 3 was that in level 3 ETCS would provide train integrity proving to allow the elimination of conventional trackside train detection equipment and allow moving block operation.

In the most recent set of specifications (Baseline 4), level 3 was merged with level 2, so that level 2 now covers all radio-based solutions, with various options for train tracking, train integrity and the use, or not, of lineside signals. Level 3 is therefore an obsolete term. However, to aid understanding, particularly with the number of historic references that exist to levels 2 and 3, the three original concepts are still explained below.

Level 1: Intermittent data transmission using balises (fixed block operation)

In brief: ETCS level 1 provides in-cab signalling that is superimposed on conventional lineside signalling to help the driver to drive both safely and efficiently. It also provides automatic train protection functionality.

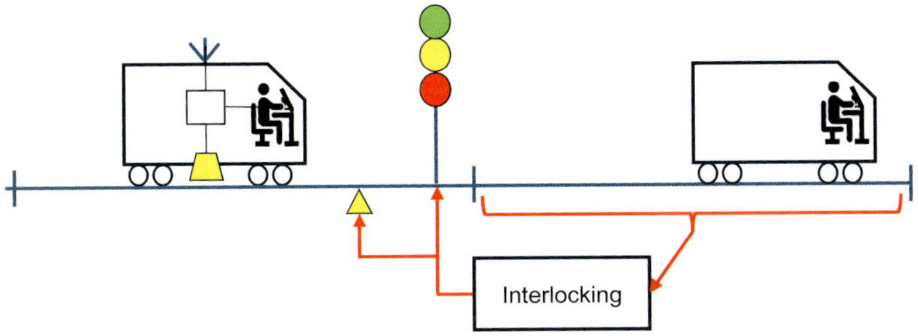

Figure 1-1: ETCS level 1

Communication from track to train is provided by 'controlled Eurobalises' which are placed on the track on the approach to lineside signals. They are connected to the signalling system to gather information on the status of signals, and they communicate the information using localised transmission as the train passes over each one. Balises are 'tele-powered' by a radio frequency signal from the train, so no lineside power supply is required. Cables link the lineside signalling to each balise, which carry the data that is to be transmitted by the balise. Lineside signals are retained although they may be simplified.

Depending on how it is applied to an existing lineside system, level 1 may reduce track capacity, primarily because if a signal changes to a less restrictive state (aspect), the on-board system will continue to force the driver to drive in accordance with the more restrictive state until it is updated when the train passes over the next balise. This drawback can be mitigated by the provision of additional 'infill' balises on the approach to signals.

Although data transmission in level 1 is intermittent (at each balise), it usually provides continuous supervision of train movement along a route as each balise provides a movement authority up to or beyond the next one. There is, however, an option known as Level 1 Limited Supervision (L1LS) which uses a subset of level 1 functionality to supervise the train movement at specific locations as a replacement for a national train control system that does not provide continuous coverage.

Level 2: Continuous data transmission using radio (fixed block operation)

In brief: ETCS level 2 provides in-cab signalling and ATP without the need for lineside signals (although they can still be provided). Used in conjunction with conventional track-based train detection equipment (track circuits or axle counters), block sections are still fixed in length, but their lengths are unconstrained by signal sighting or train braking distance considerations. Block lengths can therefore be set to optimise capacity on a route more effectively than is the case with lineside signalling.

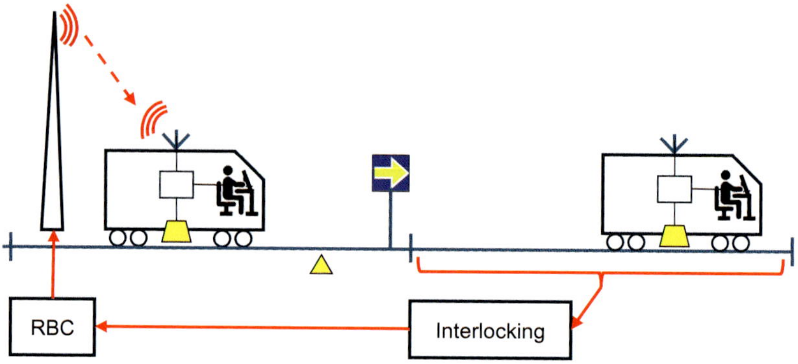

Figure 1-2: ETCS level 2 fixed block

Level 2 is a continuous ATP system where track to train communication is provided by the GSM-R radio system (although some non-European implementations use a different radio bearer, and in future FMRCS will be used). A Radio Block Centre (RBC) translates information from the trackside signalling systems into ETCS movement authorities and manages communications to and from the trains.

Radio bearer systems are inherently non-vital, meaning that the integrity of messages cannot be guaranteed by the bearer system itself. The transmission of messages in level 2 therefore uses a safe transmission protocol known as EuroRadio that virtually eliminates the risk of message corruption.

Train detection in level 2 (fixed block) can be implemented using track circuits or axle counters, and lineside signals can be either retained (for instance to allow unfitted trains to be run during a phase migration to level 2 without lineside signals), or they may be replaced by fixed marker boards which indicate the end of block sections.

Balises are used in level 2, but most are not controlled, meaning that they do not transmit variable data. They are used primarily as reference points for the train to determine its location accurately as it passes over them, enabling the correction of any on-board odometry errors and maintaining positional correlation between the on-board and trackside equipment. These passive (uncontrolled) balises are pre-programmed with the necessary positional and other data and, as with controlled balises, they are tele-powered. However, unlike controlled balises, they need no cabling for data.

Level 2: Continuous data transmission using radio (moving block operation)

In brief: Formerly known as ETCS level 3, this version of level 2 provides in-cab signalling and ATP in the same way as level 2 (fixed block). However, the key difference compared with level 2 (fixed block) is that the location of each train is reported to the trackside infrastructure using positional information sent from the trains, rather than using track-based train detection equipment.

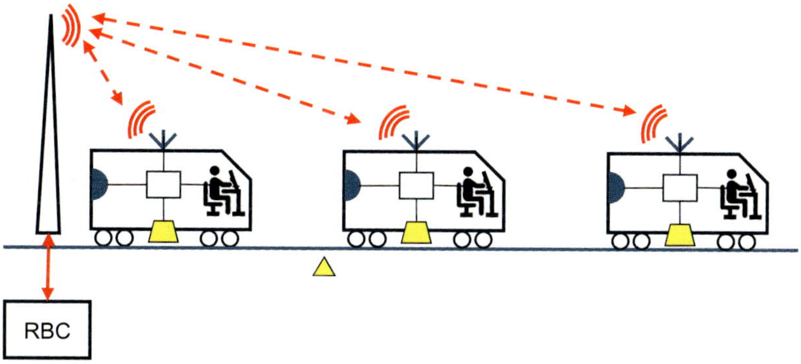

Figure 1-3: ETCS level 2 moving block (former level 3)

As with level 2 (fixed block), the radio bearer system is used for communication between the trackside and on-board systems. There are no lineside signals and no track-based train detection equipment. The ETCS on-board tracks its own position and reports it to the ETCS trackside via the radio, together with a 'train integrity' report which confirms that the rear part of the train has not become detached. Passive balises provide intermittent accurate position information to correct any on-board positioning errors, as in level 2 (fixed block).

Track capacity can be increased compared with level 2 (fixed block) because the elimination of fixed length train detection blocks removes a major constraint on the optimal use of the theoretical maximum capacity. The system can thus become 'moving block', with the distance between successive trains constrained only by braking distances plus an allowance for latency (time delays in the system) and accumulated positional errors. This should enable very short headways to be achieved, as demonstrated by communication-based train control (CBTC) systems used in metro railway applications.

A significant issue with this level 2 configuration, which has limited its adoption to date, is that the function of proving 'train integrity' becomes the responsibility of the on-board equipment, whereas with track-based train detection both parts of a train that has split are detected. For fixed-formation passenger trains, train integrity can easily be provided by a train line (linking the front and rear of the train), but for variable train formations (particularly freight trains), providing the train integrity function in a way which meets safety requirements without imposing significant operational and cost burdens has thus far proved to be a significant challenge.

1.3 Why use ETCS?

There are several potential benefits for railways that adopt ETCS, and they can be summarised under the following headings:

Interoperability – An ETCS fitted train can operate in various levels and modes to suit the operational requirements of different types of railway (e.g. dedicated high-speed lines, mixed traffic main lines, high-capacity urban lines) and can operate seamlessly across international borders. This reduces the need to install multiple types of national train control systems on a train.

Safety – ETCS includes a comprehensive automatic train protection system which fully supervises train movements, preventing collisions due to signals passed at danger (SPAD) and derailment due to over-speeding. ETCS can also reduce the need for maintenance staff to work on the track, as a consequence of the elimination of track-based equipment in ETCS level 2. It can also reduce the risks to personnel when working trackside by the provision of facilities such as temporary speed restrictions and emergency stop messages.

Performance – By eliminating the constraints of conventional lineside signals and track-based train detection, ETCS can be a facilitator for railway performance improvements such as higher speeds, increased capacity and faster recovery from service disruption.

Obsolescence – As the components used in existing national train control systems become obsolete, maintaining existing installations and obtaining equipment from the original supplier for new trains or track installations becomes increasingly difficult. Migration to ETCS (notably level 2) reduces or even eliminates the problems associated with legacy signalling systems, and the standardisation of ETCS, together with its widespread application and availability from multiple suppliers, reduces the risk of future equipment obsolescence.

Cost – ETCS is an open standard, allowing new suppliers to develop compatible products without intellectual property right (IPR) constraints. This facilitates a more competitive supplier market in which both established and new suppliers can compete. This is a marked contrast with the historical situation where there was often a single supplier of equipment for each national ATP system. ETCS level 2 also enables the elimination of equipment such as signals and track-based train detection, which reduces the cost of signalling renewals and ongoing maintenance (although of course there is more equipment on-board the trains).

In chapter 5 we revisit these benefits to review how far they have been achieved in practice.

Chapter 2: Governance and Stakeholders

2.1 Governance

The direction of, and much of the funding for, the development of ETCS and its specifications has come from the European Union (EU), as part of a wider initiative to promote rail traffic within (and between) Member States and to create an open market for the supply of railway equipment. This has included:

- Splitting the former national vertically integrated railway organisations into Infrastructure Managers (IM) and Railway Undertakings (RU) to allow the possibility of open access for competing passenger and freight services on the rail networks.
- Definition of a Common Safety Method on Risk Evaluation and Assessment (CSM-RA) to ensure a consistent and appropriate level of rigour in safety approval processes.
- Publishing Technical Specifications for Interoperability (TSI) to ensure the interoperability of new/upgraded rail infrastructure and trains across the EU.

TSIs are legally binding documents, published as 'Commission Implementing Regulations (EU)' and available in all the official languages of the EU member states. TSIs have been produced for the various elements of a railway that need to be harmonised to ensure interoperability, including infrastructure, rolling stock, traction energy systems, operations and, importantly in the context of this book, the TSI for control-command and signalling (CCS). ETCS falls within the scope of the TSI CCS.

Each TSI specifies the 'essential requirements' for various subsystems and interoperability constituents, together with approval processes for their use, to ensure and enforce interoperability. Each TSI also references a number of more detailed mandatory technical specifications.

The TSI CCS and its associated specifications are updated periodically to include additional functionality and to address technical issues that have arisen from experience in deploying earlier versions. Each update defines a 'Baseline' version of ETCS, together with the circumstances in which its use becomes mandatory for infrastructure and trains. At the publication date of this book (2026), the current version of the TSI CCS is Commission Implementing Regulation (EU) 2023/1695 of 10 August 2023, which defines ETCS Baseline 4.

Appendix A in Annex 1 of the TSI CCS lists the mandatory specifications that contain the technical requirements for ETCS. These supporting documents are all written in English and are freely accessible on the website of the European Union Agency for Railways (ERA). Archive versions corresponding to earlier baselines are also available, and in addition the ERA publishes various other documents that are not part of the mandatory standards but provide guidance and supporting information on the application of ETCS.

The ERA manages the process of maintaining and updating the TSIs and the supporting specifications in accordance with EU policy and in response to input/feedback from the stakeholders involved in the European railway industry. The European Commission has provided funding through various research and development initiatives to support the development and trialling of new functionality, most recently under the 'System Pillar' of the Europe's Rail Joint Undertaking (EU-Rail).

2.2 Stakeholders

A very wide range of stakeholders has been involved in the development and application of the ETCS specifications, and several organisations have been set up to represent their interests. Although these have evolved over time, they remain active in drafting and reviewing the detailed ETCS specifications.

Suppliers

Because ETCS is a complex technical system based on computers and software, the equipment suppliers are vital stakeholders in the technical specifications. Early in the system development they formed a suppliers' collaboration organisation known as UNISIG. The original member companies were Adtranz, Alcatel SEL, Alstom, Ansaldo, BTR Rail, CSEE Transport and Siemens. Later, as the industry changed, it comprised for many years the 'big six' – Alstom, Siemens, Invensys, Bombardier, Thales and Ansaldo. Since then, others have joined, and with the mergers and takeovers in the industry the current membership is Alstom, AŽD, CAF, Hitachi, MerMec, Progress Rail and Siemens Mobility, with Scheidt & Bachmann and Mipro as partners. UNISIG works on the basis of consensus to develop the detailed technical specifications (known as SUBSETs) which are then reviewed by other stakeholders and eventually approved and published by the ERA.

Railways

The other key stakeholders are the companies who run the railways. Within Europe they are split into:

- Infrastructure Managers (IM), who are primarily represented by the association of European Railway Infrastructure Managers (EIM).
- Railway Undertakings (RU) (the passenger and freight train operators), who are primarily represented by the Community of European Railway and Infrastructure Companies (CER).

An organisation called the EEIG ERTMS Users Group, established by several of the larger IMs, is also an official partner with ERA and UNISIG in publishing the technical specifications.

Approvals bodies

The EU legislation that has made ETCS compulsory for new and upgraded/renewed European railways also introduced a mandatory approvals framework. The purpose of this was to prevent any individual railway from requiring the use of its own approvals process, which might then become a barrier to entry for cross border traffic.

The legislation requires that independent accredited assessment bodies (known as Notified Bodies or NoBos) must perform conformity assessments of both the component parts of ETCS (known as interoperability constituents) before they can be placed on the market by a supplier, and also individual railway projects (specific applications of ETCS, whether trackside or on-board) before they can be placed into service.

In each member state there are Notified National Technical Rules and associated Technical Specifications which address issues such as the use of older (non-ETCS) train control and protection systems. Elements of railway projects that make use of these rules and specifications within a member state are assessed for conformity by Designated Bodies (DeBos), who are appointed by the national authorities.

The term Assessment Body (AsBo) may also be encountered in connection with the application of ETCS. These are independent bodies which verify that the Common Safety Method for Risk Evaluation and Assessment (CSM) has been properly applied to a railway project. Use of this CSM is a requirement for any ETCS project implemented in Europe, and often outside Europe as well.

The final approval for entry into service of an ETCS project is the responsibility of the designated National Safety Authority in each member state.

Organisations outside of the European Union

The legal requirements of the TSIs, including the TSI CCS, do not apply in non-EU countries, but to gain the benefits of ETCS standardisation the detailed specifications may be adopted as standards by other countries, together with elements of the approval regime such as the conformity assessment of interoperable constituents.

The EEIG ERTMS Users Group includes representatives of IMs in countries adjacent to the EU, such as Norway, Switzerland and United Kingdom, but for other railways outside Europe their means of influencing the further development of the ETCS specifications is largely dependent on their relationships with their suppliers, who in many cases (but not all) are also suppliers to the EU railways and are therefore involved in organisations such as UNISIG.

Chapter 3: Evolution of ETCS Specifications

3.1 Baseline 1

The specifications that define and underpin ETCS have been evolving for many years, and there has been a continual process of updates to refine them and introduce additional functionality. The set of specifications that define ETCS at a particular point in time are referred to as 'Baselines'. In this chapter we provide an overview of the development of the ETCS standards from first conception to Baseline 4, which is the latest at the time of publication of this textbook (2026).

ETCS began as an initiative of the International Union of Railways (UIC) in 1990 in the form of a new specification for automatic train protection (ATP) systems. However, it was soon adopted by the European Union with the political objective of rail interoperability within Europe. Initially, the focus was on the high-speed passenger network but was subsequently broadened to include conventional lines.

The key documents that were first produced to define ETCS were the Functional Requirements Specification (FRS), which defined the high-level requirements of the railways, and the System Requirements Specification (SRS), which defined the detail of how the functions were to be implemented in a way that would deliver interoperability. These are now supported by a large number of additional specifications, including Functional Interface Specifications (FIS) and Form Fit Function Interface Specifications (FFFIS).

The first version of the FRS was published by the European Rail Research Institute (ERRI) in 1996 (ERRI was at that time the UIC's research organisation), and it was focused on high-speed lines. Following concerns about the pace of progress, a number of other organisations were formed in order to assist the work. Three of the railways (SNCF, DB and FS) formed a user group to test out and help refine the specifications, and they were soon joined by three more (RENFE, NS and Railtrack). As a consequence, SRS V.5.a was published in July 1998.

The supply industry formed another group, initially called EUROSIG but soon renamed UNISIG. This group took over the task of further developing the SRS, correcting and enhancing the 1998 baseline specification, leading in 1999 to the first publication of the SRS in its current format, known as SUBSET-026. This version of the SRS (1.0.0) later became referred to as Baseline 1.

Meanwhile, the overall development of the suite of Technical Specifications for Interoperability, of which ETCS would form a key part, was being managed by the Association Européenne pour l'Interopérabilité Ferroviaire (European Association for Railway Interoperability, AEIF), a legal entity jointly owned and managed by the railways and industry.

3.2 Baseline 2

ETCS SRS version 2.2.2 was published in 2002 and was subsequently accepted by the European Commission as mandatory for both high-speed and conventional railways.

Between 2000 and 2006, the railway companies developed various further requirements, as a result of Directive 2001/16/EC which extended ETCS to conventional railways. This extension of the interoperability requirements added significant complexity.

With the formation of the European Union Agency for Railways (ERA) between 2004 and 2006, responsibility for management of the ETCS specifications moved to the ERA. Most of the operational staff of the AEIF transferred to the ERA and the AEIF was dissolved.

When the ERA took over responsibility in 2006, the ETCS specifications contained many optional elements (most of them addressing the needs of individual member states), which had the effect of limiting interoperability. The specifications were revised in the following year, leading to the publication of SRS version 2.3.0. Using this version, a number of railway operators started to deploy ETCS on a large scale.

Following experience of railway operation using this version, the ERA published a set of change requests in the document SUBSET-108 which defined a modified SRS (version 2.3.0d – 'debugged'). This was made mandatory by the EU in a Commission Decision modifying the high-speed and conventional rail TSIs in 2008. This version of the SRS was widely applied, and was also used for the development of the Chinese Train Control System (CTCS) for the high-speed railway network in China.

3.3 Baseline 3

SRS 2.3.0d still had shortcomings, and in 2012 the ERA commenced a significant rewrite that led to SRS version 3.0.0. This continued the evolutionary process, adding features intended to make ETCS easier to deploy on conventional lines, including level crossing functions, a standardised braking model (see chapter 12) and additional operating modes such as limited supervision and passive shunting (see chapter 8). Whilst this was undoubtedly beneficial in the longer term, it involved significant work for the suppliers in modifying their designs and products, which led to further delays in the system roll-out.

It was at this stage that the concept of baselines was formally introduced, with the previous version defined by SRS 2.3.0d becoming Baseline 2, and the new SRS 3.0.0 becoming Baseline 3. Baseline 3 introduced requirements for backward compatibility to ensure that new trains with Baseline 3 ETCS on-board could operate on trackside infrastructure compliant with Baseline 2 (see chapter 14 for more about management of system versions).

Further updates resulted in Baseline 3 Maintenance Release 1 (MR1) in 2014 (SRS version 3.4.0), which corrected various errors, and Baseline 3 Release 2 (R2) in 2016 (SRS version 3.6.0), which included the option to use packet-switched data (GPRS) for the radio communications via GSM-R in ETCS level 2. This feature was an important enabler for the use of ETCS in complex terminal and interchange stations where there were insufficient circuit-switched channels available within the allocated GSM-R frequency band to support ETCS operation (see chapter 18).

Although all Baseline 3 ETCS on-board systems are compatible with Baseline 2 ETCS trackside systems, the converse is not true, and consequentially the placing into service of new applications of Baseline 2 ETCS on-board systems is no longer permitted in the EU.

At the publication date of this book (2026), Baseline 3 (R2) is the version of ETCS that is being deployed most widely on projects around the world.

3.4 Baseline 4

The latest release (in September 2023) is Baseline 4 (SRS 4.0.0). As well as correcting various errors and amalgamating ETCS levels 2 and 3 into a more comprehensive level 2 as described in chapter 1, Baseline 4 includes or facilitates some significant enhancements such as Automatic Train Operation (ATO) and migration from GSM-R to FRMCS. However, not all of this new functionality has yet been fully validated in practical trials, and a further update is anticipated in the next version of the TSI CCS in 2027 or later, when those trials have been satisfactorily completed.

3.5 Variants outside Europe

ETCS has become something of a world standard, but in areas outside of the European Union where ETCS is not mandated through legislation, some divergences from the baselines have occurred. For instance, in China the CTCS3 system used on high-speed lines is based on Baseline 2 but includes features to provide backward compatibility to a legacy ATP system using coded track circuits. CTCS3 has been immensely successful and has accumulated more kilometres of operational running than all the other ETCS systems in the world put together, which has given the suppliers useful experience that has led to improved reliability of their equipment.

Another area of divergence is variation in the radio systems used for ETCS level 2. In some parts of the world the radio frequencies required for GSM-R are not available, and elsewhere railways are now reluctant to adopt GSM-R because the underlying 2G technology is regarded as approaching obsolescence. To address this issue, some suppliers have adapted ETCS to make use of TETRA or LTE (4G) radio systems.

Chapter 4: Game Changers and Developments

4.1 Game changers

Since the first edition of this book was published, a number of initiatives and additional features have proposed for ETCS. In many cases, these have been developed to enhance performance and thus improve the business case for ETCS. In particular, in the late 2010s, the European Commission together with the infrastructure managers (IM), railway undertakings (RU) and the supply industry decided to improve ETCS capabilities by adding a range of functions which became known as the 'game changers'. Funding for this work has been provided through the European Union's Shift2Rail research project and the Europe's Rail Joint Undertaking. The definition of what is a 'game changer' has been somewhat fluid, but includes:

- Automatic Train Operation (ATO)
- ETCS moving block without track-based train detection (formerly called level 3)
- Next generation communications to support increased functionality
- Improved train positioning (including the use of satellites)
- Harmonised operational concepts
- Braking curves optimisation

4.2 Automatic Train Operation (ATO)

ATO or GoA2 (Grade of Automation 2) is a level of automation whereby under normal conditions the train is driven by an automatic system. A driver is still retained, partly as a backup in case of operational problems, and to perform functions such as supervising train door closure at a station.

ATO and even higher grades of automation such as UTO (Unattended Train Operation – GoA4) have been common in communications-based train control (CBTC) systems used in metros for many years. The main benefit is increased capacity, because the ATO can drive much more consistently than a human. It can also save energy by driving the train more optimally. UTO has the further advantage of eliminating driver costs and allowing a more flexible service to be delivered. For instance, additional trains can be brought into service to support increases in demand more rapidly, as there are no issues relating to driver rostering.

For mainline railways, the implementation of such automation presents additional challenges. Metros are, or can be, 'closed' systems with risks such as passenger incursion onto tracks controlled by virtue of being in tunnels or by fencing and platform screen/edge doors. On a mainline railway the scale of a network makes provision of such protection much more difficult, and the existence of features such as level crossings introduce additional hazards. Nevertheless, as far back as the 1990s in the UK, it was realised that for projects such as Thameslink and Crossrail delivery of the desired headway through the central London sections of the routes using manually driven trains would be all but impossible due to the signal spacing, sighting and gradient issues. For Crossrail the decision was made to use a proven CBTC for the core section, but Thameslink opted for ETCS and became the first commercial implementation of ATO over ETCS level 2. As there were no standards for the ATO/ETCS interface this was of necessity a bespoke solution from a single supplier, and the goal of the 'game changer' development was then to define the standards required for an interoperable ATO over ETCS.

There is considerable pressure to implement ATO more widely on mainline railways, as the improvement in driving consistency reduces conflicts between trains, thus releasing additional capacity. The 'smoothing' of the train running pattern also means that energy is not wasted accelerating the train to an unnecessarily high speed only to have to brake, perhaps to a stand, because a station platform remains occupied by an earlier service or because a junction has not yet cleared.

Technically, ATO on mainline railways is now wholly possible, but there are significant issues around maintaining driver competence for manual driving, sustaining driver attention for other tasks when their primary task is done by the ATO, and managing the risks associated with transitions into and out of ATO areas.

These problems, together with the potential industrial relations issues around changing the driver's role, and the desire to gain some of the benefits of ATO in areas where a full ATP system (such as higher level ETCS) is not yet fitted, has led to the development of secondary systems to advise the driver as to the optimal speed at which to drive. These are called Driver Advisory Systems (DAS) and are on-board systems which monitor the progress of each train against the timetable. Enhanced forms of DAS, known as Connected Driver Advisory Systems (C-DAS), can communicate with the control centre and thereby receive more accurate information about the optimal driving speed which takes account of events on the network, such as the knock-on effects of delays to other trains.

The 2023 TSI CCS ETCS Baseline 4 now includes the functionality and interfaces to work with an interoperable ATO, and a description is given in chapter 13 of this book. This includes an option for display of DAS information on the ETCS DMI. Detailed specifications for the ATO Baseline 1 functionality are also included in the TSI CCS, and the aim is to validate these in time for the next TSI revision.

4.3 ETCS moving block without track-based train detection

The ETCS specifications have always included an option for trains to report their position to the ETCS trackside without the need for conventional trackside train detection (track circuits or axle counters), thus making possible 'moving block' functionality and improving headways. Originally this was called ETCS level 3, but it has now been assimilated into level 2.

Implementation of this feature has been very slow for a variety of reasons. ETCS started its life as a high-speed line signalling system, and on high-speed lines any capacity improvements gained by shortening the block lengths, including moving block operation, are marginal because the braking curve for a high-speed train is so much longer than the block lengths. It is only on high density commuter type lines operated at lower speeds where moving block operation can deliver significant capacity benefits. However, these lines are by their nature heavily used and any disruption is therefore highly undesirable; they are not generally the place to be experimenting with new signalling technologies. The ETCS level 3 pilots that have been implemented have therefore tended to be on low density rural routes where the operational and reputational risks are much lower. Whilst these pilot installations have been able to test some features of the system, the low density of the lines on which they have been installed means that the system has not really been stress tested.

Another obstacle to implementing ETCS level 3 on conventional routes is the 'train integrity' function for freight and other trains that do not have a fixed formation. In a fixed formation train it is relatively easy to provide this function on-board via the train lines that run the length of each train, but for freight trains where wagons are attached, detached and re-ordered on a regular basis, such solutions are impractical. On-board 'end of train' devices exist but it is proving difficult to achieve a validated safety integrity greater than SIL2. Although this may be sufficient for some freight-only lines, it is insufficient for mixed traffic lines (i.e., carrying both passenger and freight traffic).

The emerging solution is the development of a Digital Automatic Coupler (DAC) for freight trains. The DAC is an initiative aimed at improving rail freight efficiency in Europe which, amongst other benefits, will provide digital communication along a freight train for a wide range of functions including train integrity. A standardised specification for DAC is under development and trials and will be specified in a future TSI.

Another significant achievement of the game changer initiative has been the demonstration of a 'Hybrid Level 3' solution, where ETCS moving block is overlaid on conventional signalling. This allows moving block operation for an ETCS fitted train which is following another train that has train integrity proving/reporting, whilst retaining fixed block operation using track circuits or axle counters to detect the location of trains that do not have integrity proving.

From the point of view of the ETCS on-board, the only difference between ETCS level 2 and what was originally called level 3 was the provision of the train integrity input. The movement authority to a train is the same whether it has been calculated based on a train position report from another train or on track-based train detection equipment. In the 2023 version of the TSI CCS this was clarified by abolishing level 3 as a separate level, and so ETCS level 2 now fully supports the use of train-based position reporting, both for routes and networks where all trains have this capability, and as part of a hybrid solution using both track- and train-based position reporting (see chapter 7).

4.4 Next generation communications to support increased functionality

The existing GSM-R communications system, which form part of the ERTMS specification suite, was formulated by the UIC EIRENE project in the mid-1990s and is based on 2G mobile radio technology. As early as the late 1990s on the West Coast Route Modernisation project in the UK, it was realised that the number of circuit-switched channels (GSM radio frequency pairs) allocated for railway use would be insufficient to support urban rail operation in ETCS level 2. The specific problem was that the number of trains within a radio cell area could easily exceed the number of circuit-switched channels available. It was concluded that the use of GPRS packet switched data could solve this problem, as this allows much more efficient use of the available radio bandwidth and enables many more trains to communicate within each radio cell. However, the use of GPRS did not appear in the specifications until Baseline 3 Release 2.

Meanwhile, the commercial world of radio continues to move on, with 5G now implemented by public mobile networks in many places and 6G under discussion. The railways have done very well in persuading equipment suppliers to continue to support GSM-R and thus defer major obsolescence problems, but that cannot last forever and data communications requirements on railways are still increasing. The limited capacity of GSM-R, even using GPRS, has meant that it only has enough capacity for operational voice communications and ETCS data and, partly on account of this, railway operators make extensive use of public networks for other purposes.

Standards bring many benefits and are essential for interoperability, but they can also prevent innovation. The sheer scale, both numerically and geographically, of the communications investment required for railways makes the replacement of GSM-R a major challenge, and the new system will need to co-exist with GSM-R through a transitional period. The UIC initiated a project called 'Future Railway Mobile Communication System' (FRMCS) which has led consultations on the options for replacing GSM-R (which included the option of using commercial networks with the appropriate service level and security protections) and has developed specifications for FRMCS. At the time of writing (2026), several trial implementations of FRMCS are underway, involving different suppliers to validate and refine the draft specifications, and to demonstrate the much wider range of railway applications that will be enabled by the increased capacity that FRMCS offers.

The 2023 TSI CCS includes preliminary versions of the FRMCS standards and specifies how the EuroRadio interface between ETCS and GSM-R needs to be enhanced to interface additionally with FRMCS (including additional cyber-security measures), as part of Baseline 4 for the ETCS on-board (see chapter 18). This has proved controversial as it implies that all existing trains fitted with ETCS Baseline 3 and its subsequent maintenance releases will require an update to Baseline 4 as part of the FRMCS roll-out.

4.5 Improved train positioning (including with satellites)

A key benefit of ETCS level 2 is the potential reduction in lineside equipment (signals and track-based train detection systems). As well as reducing capital costs, there is an ongoing cost and safety benefit by dramatically reducing the need for personnel to go onto the track to maintain signalling equipment. At present, even with the elimination of lineside signals and train detection, ETCS still requires track-mounted balises to enable trains to verify their position, and eliminating these could lead to further financial and safety benefits.

Pilot studies are underway to determine how to use Global Navigation Satellite Systems (GNSS) as an ETCS train location method, without the need for balises. Historically, the dominant concern about GNSS for railway applications was the fact that there was only the Global Positioning System (GPS) available to use, which is owned and controlled by the USA. However, several other GNSS systems are now in operation, notably Galileo (European Union) and GLONASS (Russia). Most GNSS receivers can use multiple GNSS satellite systems, thereby enhancing availability. GNSS is already used extensively in other transport sectors, as well as for railway applications such as driver advisory and passenger information systems. It is also used by some Positive Train Control (PTC) systems in North America.

The concept behind the use of GNSS for ETCS is to replace physical balises with 'virtual balises'. To enable this functionality, the ETCS on-board would have to include a new module called a Virtual Balise Reader (VBR). The VBR includes a GNSS receiver which processes the GNSS coordinates to determine the location at which the train passes a virtual balise. As the train passes the virtual balise location, the VBR generates the corresponding balise telegram for the ETCS on-board. This approach means that minimal changes to the ETCS on-board are required.

Several challenges must be addressed before this solution can be adopted and incorporated into the ETCS standards. These include providing positional accuracy that is sufficient to identify which track a train is on, coverage in locations such as tunnels, and management of the on-board route map that is used for matching GNSS coordinates to virtual balise locations. The risk of malicious interference (as has occurred in the air sector) also needs to be assessed and mitigated.

Another area of interest for alternative position-sensing technologies is to replace, or reduce the dependence on, on-board odometry. Odometry errors lead to increasing positional uncertainty as a train travels. In ETCS these errors are removed periodically as the train passes over the track-mounted balises, which provide accurate position references. The greater the rate at which positional errors increase, the more balises are needed to remove them, leading to a system level 'trade off'. Speed is traditionally measured on-board using axle odometers which count wheel rotations. Some errors are caused by progressive wheel wear, but the most significant errors occur when slip/slide occurs in poor adhesion conditions. Various methods are used to minimise that source of error. Typically, two odometers are used, mounted on different axles, ideally on unbraked and unpowered axles so that there is the minimum likelihood of slip/slide. Alternatively, a secondary sensor that does not rely on axle rotation may be used, such as a doppler radar. Even with these arrangements, odometry errors can still be an issue and work on various alternative methods, including magnetic and laser sensors and video image processing, has continued. There is no barrier to these being used in an ETCS solution, as there is no interoperable interface to be respected. More about odometry can be found in chapter 17.

This 'game changer' remains very much a 'work in progress' but may eventually be covered in a future version of the TSI CCS.

4.6 Harmonised operational concepts

Much of the complexity of ETCS comes from the need to accommodate the different operational rules and signalling principles in the countries where it is used. Each country's operating rules have developed over decades and as a consequence there are very significant differences, even within Europe. Harmonising the rules would make the movement of trains across international borders more straightforward, and could reduce ETCS system complexity. The costs and risks associated with drivers and others learning the operating rules of more than one country are substantial. However, by using ETCS level 2 without lineside signals, some of the challenges are eliminated (for instance, there are no lineside signal aspects and aspect sequences to learn, and route knowledge is less of an issue).

This is yet another topic which has been under discussion for many years without, as yet, much resolution.

4.7 Braking curves optimisation

The ETCS braking model (see chapter 12) makes conservative assumptions that result in the supervision limits being further from the target than is necessary. Some improvements have been made in the Baseline 4. For example, there is new functionality to use reduced brake build-up times if the ETCS on-board can determine that a partial brake application is sufficiently strong enough to bring the train speed below the target.

There are other optimisations that are currently under consideration but not yet implemented. For instance, as part of the calculation of its supervision limits the ETCS on-board considers that the train will coast (and possibly accelerate depending on the parameters) during the emergency brake build-up time. But if the train is already braking, a transition from braking to coasting/acceleration is impossible in such a short calculation interval.

Looking further into the future there are ideas that could result in further optimisations, such as feedback on the health of the train's braking system and feedback on the degree of wheel slip/slide activity which could both be used to dynamically adjust parameters with the ETCS braking model.

Chapter 5: ETCS today – What has been achieved?

5.1 Deployment of ETCS

By the end of 2024, over 100,000 km of track and 30,000 train cabs were in operation or under contract to be delivered in countries around the world. While ETCS set out to be a European standard, there are now applications on every continent where railways exist.

In this textbook we are not attempting to describe any of the specific projects that have deployed ETCS. A companion volume 'ETCS in Europe' will describe the approaches to implementing ETCS in several European countries, and there have been many articles describing specific projects in the Institution of Railway Signal Engineering monthly journal IRSE News.

However, it is worthwhile to mention here the following projects, all of which have made a significant contribution to the development of ETCS.

- The Mattstetten to Rothrist line in Switzerland was equipped with ETCS using a very early version of the specifications (V.4.a). It demonstrated that a level 2 system could work, and at the same time exposed a significant number of track-train interface issues. It further demonstrated that an RBC and an interlocking from different suppliers could communicate with each other.
- The decision in Spain to adopt ETCS level 2 from multiple suppliers for the country's high-speed network exposed many issues.
- China chose ETCS (at Baseline 2) as the basis of their CTCS3 cab signalling system. This has been a major success, with over 40,000 km now implemented across the country, the vast majority on high-speed lines. Although not a central part of ETCS implementation as conceived in Europe, it has contributed many reliability improvements, including non-software issues such as preventing antenna water ingress at 380 km/h.
- The Vasterdalbanan project in Sweden demonstrated that level 3 (as it was then called, now level 2 without track-based train detection) was capable of working on a low-density rural railway. It also showed that RBC and interlocking functionality could be run on a single computer platform.
- The high-speed line from Antwerp in Belgium to Amsterdam in the Netherlands exposed and resolved RBC to RBC interface issues between suppliers at a border crossing.
- The Thameslink project in London showed that ATO can be successfully implemented over ETCS level 2.

5.2 Maturity of the specifications

ETCS has matured considerably since the first edition of this book, and it can be argued that every project has contributed to this progress to some degree. The manner in which the system was specified and implemented means that the improvements have fallen into three broad areas:

- Evolution of the specifications themselves. This has been driven partly by the decision to extend interoperability from high-speed to conventional rail, and by the inclusion of new functionality (for instance ATO), and partly by the correction of errors (and more recently by obsolescence issues). Conventional rail routes have many additional features such as level crossings which have required amendments and additions to the specifications.

- Correction of differences in the interpretation of the specifications. These differences had led in some cases to full interoperability not being achieved. In the early days this affected mainly the train-track interface, but later, as implementation contracts delivered by different suppliers started to interface with each other, problems were encountered in other areas, such as RBC to RBC communications. Some problems were corrected by suppliers changing their implementation arrangements, whilst others led to documented clarifications to the ETCS specifications.
- Resolving issues that were unique to individual suppliers, such as software bugs or other performance issues which sometimes only emerged when systems were in service, despite rigorous pre-service testing.

National implementation strategies have also been a major area of discussion, and have been the cause of both delay and change, as well as progress. Interoperability was a European political imperative that was (and still is) intended to improve railway competitiveness, including by making cross-border movements more straightforward within the EU. However, the harmonisation of Member States' operational rules has proved to be too great a challenge in the short to medium term, given the many thousands of personnel who need to be re-trained and the transitional risks that are associated with major change. As a result, ETCS is a very complex system, not least to enable compatibility with the operational requirements and national systems of individual Member States.

The EU also mandated the separation of Railway Undertakings (RU) from Infrastructure Managers (IM). This was to improve competition, particularly in the freight area where historically some state railways had operated a monopoly. However, this introduced an additional difficultly in producing business cases for ETCS. Replacing traditional lineside signals with cab signalling means that costs increase for the RU, whereas the benefits in terms of reduced lineside infrastructure and thus maintenance costs accrue to the IM. Furthermore, the major benefits of ETCS can only be realised (or fully realised) when all trains on a network are fitted with ETCS, and the retrofitting of older trains and locomotives has proved particularly expensive. Thus mechanisms have had to be found to harmonise costs and benefits between RUs and IMs. Some countries (e.g. Denmark and Belgium) have had the benefit of a government decision to go for a 'whole country' strategy with adequate funding to plan all of the implementation. Others have been constrained by phased funding and having to produce a business case for each discrete funded stage.

5.3 Have the benefits been realised?

At the end of chapter 1 we listed the benefits anticipated from the ETCS development. Here we consider the extent to which these benefits have been realised in practice. It should be noted that these are the opinions of the authors, not the result of an objective analysis.

Interoperability: ETCS has delivered technical interoperability between track and train, and has demonstrated its applicability to different types of railway. This has supported the development of new international passenger and freight services. However, in most cases ETCS has had to be installed on the international trains alongside the existing national train control systems for each country in which the trains are required to operate. The anticipated reduction in the number of on-board systems will not be realised until a lengthy migration period has elapsed.

Safety: The safety functionality specified for ETCS has been delivered, and as far as the authors are aware, there have been no significant train accidents as a result of any deficiency. This contrasts with the number of accidents and near misses that continue to occur with some existing national systems that provide a less comprehensive level of train protection. ETCS level 1 has been adopted by a number of railways that previously had very basic protection systems (or none at all) to deliver a significant safety improvement.

Performance: The performance impact of ETCS has been the cause of much debate, with concerns that ETCS level 1 can have a negative impact on capacity due to the intermittent communication via balises, and exaggerated claims of capacity benefits in respect of level 2. In practice, well-engineered level 1 applications have been delivered without performance issues, and the projected capacity benefits of level 2 can be realised, but only under the right conditions of train service patterns and with support of other systems such as Traffic Management, Automatic Train Operation and Driver Advisory Systems.

Obsolescence: A number of railways have seized the opportunity to use ETCS to replace an obsolete national train protection system. In other cases, it has been possible to extend the life of an existing national system for which new on-board equipment is unavailable by developing a specific transmission module (STM) that will allow the ETCS on-board to communicate with the existing national system's trackside equipment. However, obsolescence of earlier versions of ETCS is now becoming an issue. Railways that pioneered ETCS in the earliest projects have incurred expensive upgrades to bring them in line with the later baselines, and migration from GSM-R to FRMCS will be a significant challenge for the thousands of trains and kilometres of track that are already fitted with ETCS.

Cost: The complexity of ETCS means that suppliers have had to invest large sums of money in developing their products, which is inevitably reflected in the price. In addition, the application engineering to design and install the systems on both trains and on trackside infrastructure is complex, and the training of drivers and other personnel is yet another significant cost. For applications that overlay ETCS onto existing trackside systems and trains, there is an inevitable increase in cost which has had to be justified by the other benefits listed above. On the other hand, many infrastructure managers are finding that the cost of renewing conventional trackside signalling is unaffordable and are adopting ETCS level 2 without lineside signals for this reason. Retrofitting ETCS on existing trains remains very expensive, but virtually all new trains are now designed with ETCS as a standard fitment, which is a marginal additional cost.

To conclude, whilst progress has been slower than everyone had hoped for in the 1990s, gradually a system has emerged that can now be considered to be a world standard, not just a European one. ETCS is without doubt a highly usable and configurable cab signalling system, and most of the implementation issues are now sufficiently well understood to be overcome. It can be argued that for a new railway almost anywhere in the world, ETCS has become the default option except where a metro CBTC system is more applicable. The latest Baseline 4 provides a platform for further developments such as ATO and FRMCS, which will extend its applicability even further.

References

UNIFE ERTMS website https://www.ertms.net/ provides world ETCS deployment statistics.

IRSE Knowledge Base https://irse.info/kbase is a searchable library including many articles on ETCS projects in IRSE News and elsewhere.

Part 2
ETCS Concepts and Functionality

Part 2 begins with defining the role of ETCS within the wider world of railway control-command and signalling systems. The key concepts of ETCS levels and modes are described in detail, together with features such as movement authority and various types of speed supervision. The interaction with braking systems and automatic train operation is covered, together with the impact on performance measures such as railway line capacity. The concepts of standards baselines, system versions, and national values and how they are managed is explained.

Chapter 6: ETCS Context and Architecture

6.1 ETCS system context

It is important to understand that ETCS is not a complete solution for railway control command and signalling (CCS). Its role is to provide an interoperable means for communicating a movement authority from the trackside CCS to the train, to present this information to the train driver and supervise the safe movement of the train. ETCS sits alongside other elements of ERTMS such as RMR (Railway Mobile Radio) and ATO (Automatic Train Operation).

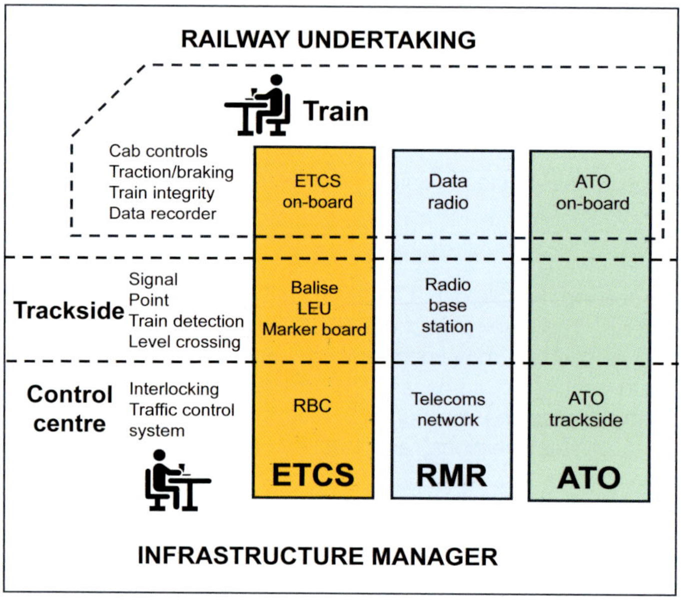

Figure 6-1: ETCS in the railway CCS context

6.2 Traditional trackside control, command and signalling

The trackside and control centre elements of ETCS sit alongside the traditional components of a control command and signalling system, these being principally the following:

Traffic control system (may also be referred to as traffic management, signalling control or centralised train control system):

- Provides the human interface to operators in the control centre (they may also be referred to as signallers, dispatchers, or controllers)
- Displays the status of trackside equipment, the location of trains and descriptive codes to denote the type, routing and destination of each train
- Allows route setting for trains by the operator or automatic route setting (ARS) in accordance with the pre-planned timetable or a dynamically updated plan and sends real time data on train movements to customer and staff information systems

Interlocking:

- Collates information from trackside equipment and reports the status to traffic control system
- Receives route setting requests from the traffic control system and checks for conflicts with train movements and other safety constraints
- Sets and locks routes and commands points to move to establish the required path through the railway network, subject to the above safety checks having been passed
- Determines the aspect to be displayed on each lineside signal, taking into account route locking, points detection and track occupation
- Releases routes after the passage of a train, to allow other trains to use the vacated track

Trackside equipment:

- Lineside signals that provide a visual movement authority to train drivers
- Points (switches) that steer the trains through the track network
- Train detection systems (track circuits or axle counters) that indicate whether or not a section of track is occupied by a train
- Level crossing equipment that manages the movement of road vehicles and pedestrians over the railway at level crossings

Note that ETCS level 2 allows the possibility of eliminating two of the trackside equipment types:

- Lineside signals, by providing cab signalling
- Track based train detection systems, by having the trains report their location and integrity by radio

6.3 ETCS and radio trackside sub-systems

The trackside elements of ETCS interface with these traditional CCS components to provide data communication to each train in parallel with, or instead of, the visual movement authority conveyed by lineside signals to the driver. They also allow each train to determine its location on the track.

The trackside elements of ETCS are:

- Eurobalises that provide an accurate position reference on the track and can also send a data telegram to the train
- Lineside electronic units (LEU) that provide an interface to allow the data telegram from a balise to be updated with information on the state of the interlocking and lineside signalling equipment (this is used in ETCS level 1 to send movement authorities to the train)
- Marker boards that advise the train driver of locations on the track that denote the precise end of a movement authority should a train be required to stop there (these may be provided in ETCS level 2 instead of lineside signals)

The control centre element of ETCS in ETCS level 2 is the Radio Block Centre (RBC), which performs the following functions:

- Registers and tracks the status and location of ETCS fitted trains in the control area
- Receives information from the interlocking, translates this into ETCS movement authorities and sends these to the relevant trains
- Sends information on the ETCS status of trains to the interlocking and traffic control system; this may include location of the trains as a supplement or substitute for lineside train detection
- Receives information from the traffic control system or another operator interface to allow the application and removal of temporary speed restrictions and other ETCS functions that are not supported by the interlocking

The RBC communicates with the train using the railway mobile radio (RMR) system, which comprises:

- A telecommunications network that links the control centre to radio base stations alongside the railway tracks
- Radio base stations that exchange messages with the voice and data radios on trains in their vicinity

The original standard radio system for RMR in European standards is GSM-R, but this will migrate to FRMCS (Future Rail Mobile Communications System) in the future. Some ETCS applications outside Europe have used other radio systems such as LTE or Tetra. See chapter 18 for more detail on the use of RMR with ETCS.

Although not shown on figure 6-1, a key management centre is an essential support for ETCS level 2. The secure exchange of information between each RBC and the ETCS on-board relies on the use of secret data encryption keys that are used to generate and validate the message authentication codes that defend the communication link against cyber-attack. Key management centres are responsible for generating the keys and distributing them to both the RBCs and to the ETCS on-board each train.

6.4 ETCS on the train

The ETCS on-board comprises several modules which may be implemented either as separate items of hardware or as software components on a common platform. There are:

- Balise transmission – communicates with balises in the track to receive data and establish an accurate position reference for the train
- EuroRadio – interface to the train's data radio (which communicates to the RBC via RMR)
- Driver machine interface (DMI) – interface with the driver, providing movement authority, location and status information as well as allowing data input
- Odometry – interface with motion sensors on the train, providing information to be combined with the position reference from balises to maintain an accurate record of train location and speed
- Cold movement detection – interface to a system that identifies whether a fitted cab has been moved during a shutdown period; this permits automatic re-initialisation of the ETCS on-board if the train/locomotive has not moved (which means that the on-board record of the train's location is still correct)
- Brake interface – interface to the train's traction and braking systems to allow ETCS to intervene and slow the train if the driver does not control train speed within the limits of the movement authority
- Interfaces to other systems on the train – such as cab controls which impact ETCS modes of operation, and a train integrity input to allow ETCS position reports to be used instead of conventional lineside train detection systems
- Data recorder interface – providing a log of ETCS information that is sent to an on-board data recording device which stores this information for subsequent analysis to review performance or investigate an accident
- Specific transmission module – providing functionality to allow the train to operate on tracks fitted with another type of train protection system (referred to in the ETCS standards as a 'national train control' or 'Class B' system)
- ATO control function – interfaces with the optional ATO on-board allowing the train to be driven automatically

6.5 Automatic Train Operation

The specification of an interoperable ATO solution is a significant step forward in the 2023 TSI CCS. A set of standards are specified for the on-board and trackside elements of ATO, and the interface with the ETCS on-board. From Baseline 4 (system version 3.0) the ETCS standards now include all the functionality required to work alongside the interoperable ATO. This is described in detail in chapter 13.

References

SUBSET-026-02: System Requirements Specification, Chapter 2 Basic System Description

Chapter 7: ETCS Levels

7.1 Explanation of levels

There are four different ETCS levels; level 0, level 1, level 2 and level NTC, each with its own technical characteristics. An ETCS on-board system is always operating in one of the above-mentioned levels. It is the ETCS level in conjunction with the current ETCS on-board mode that determines the supervision functionality of ETCS. An in-depth description of ETCS on-board modes and functionality can be found in chapter 8.

A key feature of ETCS is that it enables the train to seamlessly switch, at speed, from one level to another. This is known as 'level transition' functionality. The level can be set by the driver on the DMI, or by the ETCS on-board receiving a level transition order from the trackside (transmitted via balise or radio). The relation between equipment and level is shown in Table 7-1.

	Level 0	Level 1	Level 2			Level NTC
Lineside Signals	Generally, yes (although level 0 is sometimes used for un-signalled areas)	Yes	Generally, no (although signals may be required for handling non-ETCS trains, this is called 'overlay')			Generally, yes (most legacy national systems have signals)
Transmission	Balise for level transitions only	Balise	Balise, Railway Mobile Radio			Balise for level transitions only
Radio Block Centre (RBC)	No	No	Yes			No
Train separation	Generally, Fixed Block	Fixed Block	Fixed Block	Virtual Block	Moving Block	Generally, Fixed Block
On-board Train Integrity	No	No	No	Yes	Yes	Generally, no

Table 7-1: Relation between level and equipment

Note that level 2 includes a choice of three different train separation principles, including moving block (previously known as level 3).

Table 7-1 shows that the ETCS on-board is capable of reading balises regardless of which level the system is operating in. This means that the ETCS on board system will act on level transition orders in all levels and keep the train location information consistent when a change of level occurs.

The level transition functionality helps to minimise operational impact when migrating from a legacy signalling system to ETCS. The train can operate in ETCS level NTC on the sections of track where legacy national train protection systems are in use, and then seamlessly transition to ETCS level 1 or 2 at the border with the new ETCS sections. Levels 1 and 2 are backward compatible, so an ETCS level 2 equipped train can run on a level 1 track.

The choice of which level and train separation options to deploy on a route is very much dependent on the relevant infrastructure manager's assessment of operational requirements and infrastructure factors, along with the long-term asset management strategy. All railway undertakings using the route should also participate in the assessment, taking into account amongst other things the version of on-board ETCS they currently have deployed on their rolling stock.

7.2 Level 0

Level 0 is used to operate ETCS equipped trains on lines not equipped with ETCS or national train protection systems. For example, level 0 is often used for depots and yards. Level 0 is also used for degraded operations, for example when the ETCS trackside equipment is out of service.

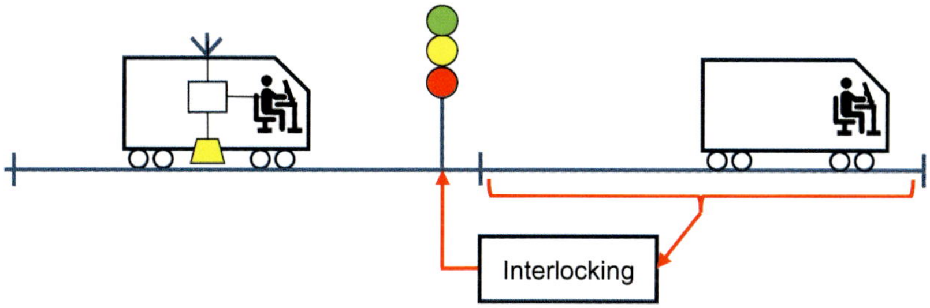

Figure 7-1: ETCS level 0

In level 0 lineside signals and signs inform the driver of movement authorities and speed restrictions. However, it is still possible to fit balises on the trackside to enable functions such as announcements for level transitions, enforcing temporary speed restrictions and text messages for the drivers.

Operationally there are only two driving modes that are used in level 0, these being UN (Unfitted) mode and SH (Shunting) mode.

- UN mode can be entered at start of mission, after the driver has completed train data entry and pressed the start button. Note that the entry of train data is required so that the ETCS on-board can transition to another level if ordered to do so later. UN mode can also be entered when transitioning into level 0 from another level. In both cases the driver must acknowledge the change by pressing the relevant button on the DMI.
- SH mode can also be entered at start of mission and has the advantage that train data entry is not required, making it very simple for depot/yard operations. The disadvantage is that, in SH mode, the ETCS on-board will not transition to another level (transition orders are stored and only evaluated when another mode is entered).

In level 0 the mode, the ETCS level and the current speed are indicated on the DMI display, but nothing else.

The ETCS on-board reads every balise group in level 0 and uses this to determine the location of the train in preparation for a transition to another level. In a transition area to level 2, the on-board will establish a communication session and report the train position to the RBC.

Balise groups can be fitted in a level 0 area as a means of enforcing temporary speed restrictions. Since there is no balise linking in the level 0 area, two balise groups with at least two balises each are needed to transmit the temporary speed restriction. The use of two balise groups is necessary to meet the safety requirements. Lineside signs or another means of informing the driver of the temporary speed restriction will also be required as in UN mode the temporary speed restriction will not be displayed on the DMI.

In level 0 the ETCS on-board undertakes speed supervision as described in chapter 11, but based on a most restrictive speed profile (MRSP) that only takes account of:

- mode related speed restriction (specified in a national value for UN and SH modes)
- train related speed restriction (the maximum speed at which the train is designed to travel)
- temporary speed restriction (transmitted via balise)
- override related speed restriction (national value applicable when 'override' function selected)

If operating in SH mode, the driver can request the speed limit for this mode to be indicated on the display by pressing anywhere on the circular speed gauge. This is known as 'toggle-on'.

7.3 Level 1

Level 1 is designed as an overlay to existing trackside infrastructure, the latter comprising interlocking or block systems, lineside signals and train detection equipment.

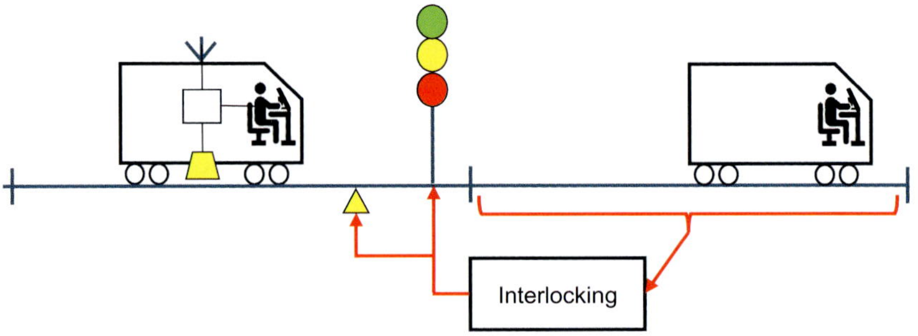

Figure 7-2: ETCS level 1

Level 1 provides continuous supervision of train speed and location. This means that all relevant speed profiles, gradients and distances are taken into account in the calculations performed by the ETCS on-board. All the information to enable driving within the supervised speed profile is presented to the driver on the DMI.

Level 1 is a spot transmission system, so called because the trackside equipment used for data transfer to the ETCS on-board is based on balises placed at intervals along the track. Static data is transferred via fixed balises (i.e. the data they transmit is unchanging), and dynamic data such as the movement authority is transferred via controlled balises. In many cases two or more balises will be installed at the minimum spacing along the track as a 'balise group'. The sequence of balises within the group encountered by a train allows the direction of movement to be deduced by the ETCS on-board.

The controlled balises receive the information regarding the current aspect of the associated lineside signal from a connection to a lineside electronic unit (LEU). The LEU receives the signal aspect information from either the signal itself or from the interlocking. It then selects the correct telegram of information for the train that is approaching the signal and transmits it to the balise ready for the train to receive when it passes over. Further information about the use of balises in level 1 is provided in chapter 16.

Note that several ETCS suppliers offer a direct connection from the interlocking to the balise, thereby removing the LEU from the architecture and instead providing balise driver modules within the interlocking itself (sometimes called a 'centralised LEU'). As well as saving hardware this method allows the interlocking to send extra information via the balise to the ETCS on-board. For example, the interlocking can be adapted to send longer movement authorities as well as information from the traffic control system such as temporary speed restrictions.

One limitation of level 1 is that, in its most basic architecture, the movement authority is only transmitted by the balise group which is located at the signal. This means that when a train has stopped on the approach to a red (stop) signal (and not passed over the balise group), the ETCS on-board cannot know when the signal changes to a proceed aspect. To deal with this, a level 1 movement authority incorporates a 'release speed'. When the driver sees the proceed aspect displayed by the lineside signal, the release speed allows the train to move forward slowly until the on-board balise reader passes over the balise group and receives the new movement authority.

This limitation of level 1 also means that if a signal changes from a red (stop) aspect to a proceed aspect while the train is moving towards it, the DMI will continue to display restricted speed or braking indications to the driver in the expectation that the train is required to stop at the signal – and of course the ETCS on-board will supervise the speed accordingly. Level 1 infill was developed to solve this problem. By installing additional 'infill' balises on the approach to the main signal, the new proceed aspect can be transmitted more promptly to the ETCS on-board. Thus, the train gets the updated information before reaching the signal, which increases the capacity of the route.

From the foregoing it can be seen that lineside signals must be retained for level 1, so that the driver of a train which is stationary at the end of a movement authority knows (from the signal aspect) when it is permissible to move forward at the release speed to obtain the new movement authority. However, the signal can be a simple stop/go indication that only needs to be read at low speeds, with less onerous safety integrity and signal sighting requirements. This has enabled some infrastructure managers to adopt a much simpler and less costly type of signal.

The original ETCS specifications included options for semi-continuous infill on the approach to a signal known as 'Euroloop' (a leaky feeder cable running along the rail) and 'Radio Infill' using GSM-R. However, in practice neither has been widely adopted and from 2020 onwards new installations of these types have no longer been permitted.

At Baseline 3 an option for Level 1 Limited Supervision (L1LS) was added to the specifications. This uses a subset of level 1 functionality to supervise the train movement at specific locations as a replacement for a national train control system. Limited information about the infrastructure is sent to the ETCS on-board at these locations, typically speed for the next section and the status of the next signal. This information is not displayed to the driver on the DMI, and the system operates in the background as a kind of train stop (see description of LS mode in chapter 8).

The key elements and characteristics of ETCS level 1 are:

Trackside equipment:

- Controlled Eurobalises to transmit variable information
- LEU or interlocking interface to send the correct telegram to the associated controlled balise, based on relevant signalling inputs (signal aspect, etc.)

Main ETCS trackside functions:

- Determine movement authorities for each train individually according to the status of the underlying signalling system
- Transmit movement authorities and track description data (gradient, speed limits, etc.) to the train

Main ETCS on-board functions:

- Receive movement authority and track description data from balises
- Select the most restrictive speed profile
- Calculate a dynamic speed profile, taking into account the distance to the end of the movement authority, the track description data and the train running/braking characteristics (which are known by the on-board systems)
- Compare the actual train speed with the permitted speed (from the speed profile), and command a brake application if necessary
- Display cab signalling information to the driver via the DMI

7.4 Level 2

Level 2 is a radio-based continuous transmission train control system. It does not require lineside signals, and the provision of track-based train detection equipment is optional. These characteristics make level 2 the preferred choice for most infrastructure managers because it improves operational performance and availability, and at the same time reduces life-cycle costs.

While the 'no-signals' architecture is the preferred end-state, level 2 is also designed so that it can be used as an 'overlay' system, whereby signals and train detection equipment are retained (similar to level 1). The level 2 overlay architecture is very useful for railways that have a mix of ETCS equipped and non-equipped trains and may be planning to migrate to an all-ETCS fitted railway in the future. Once all trains are fitted with ETCS, it is then possible to migrate from an overlay to a no-signals architecture and achieve the long-term performance, availability and cost benefits.

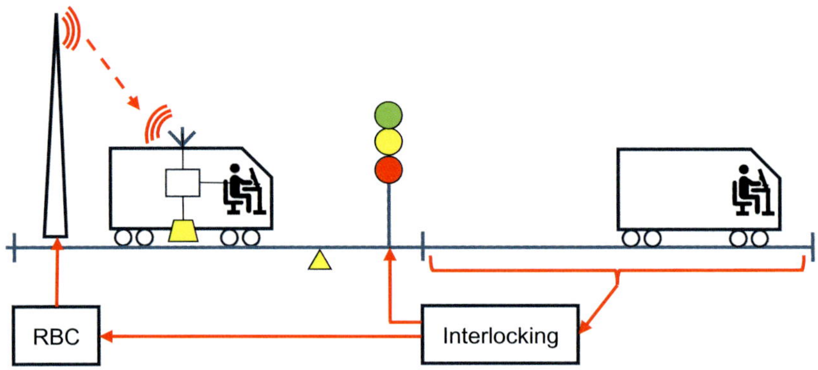

Figure 7-3: ETCS level 2 'overlay'

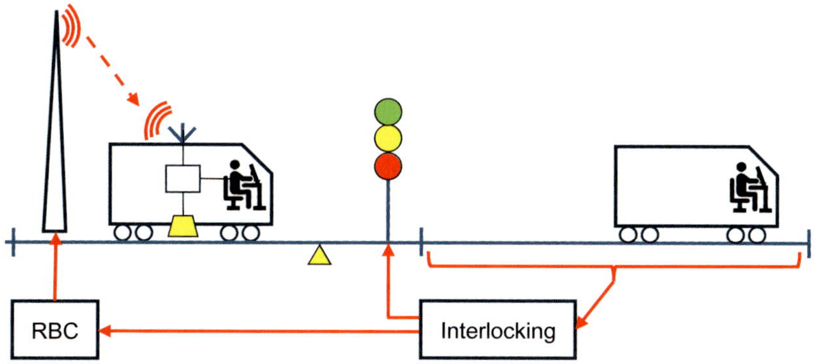

Figure 7-4: ETCS level 2 'no signals' fixed block

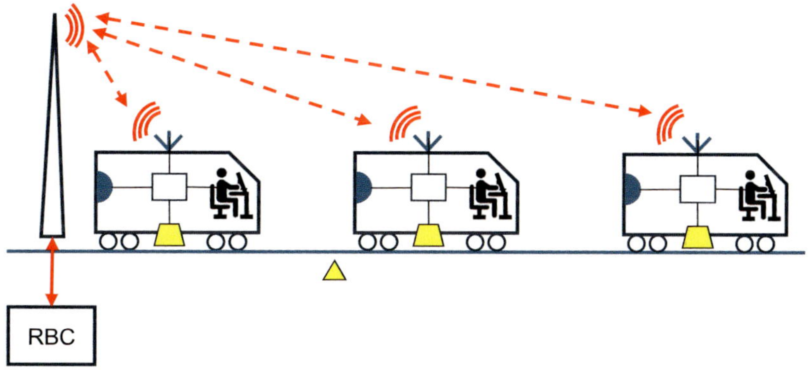

Figure 7-5: ETCS level 2 'no signals' moving block

ETCS level 2 'no-signals' trackside equipment comprises Radio Block Centres (RBCs) and fixed balises installed in the track. There may also be ETCS block markers at the trackside in place of lineside signals, to provide the train driver with a visual indication of the stopping point at the end of a movement authority.

The radio block centre interfaces to the interlocking(s). In the case of an existing railway, the interlockings are typically replaced or upgraded to support level 2. Railway Mobile Radio (RMR) enables full bi-directional communication between the ETCS on-board and the RBC. Security of the radio transmission is achieved using the EuroRadio protocol with cryptographic keys that are controlled by a key management centre (KMC) and public key infrastructure (PKI). More detail about RMR, EuroRadio and key management is provided in chapter 18.

In the RBC, all static data such as static speed profiles, gradient profiles, track conditions and the locations of balises are held in the form of a route map. Through a direct connection between the RBC and interlocking, all the necessary dynamic information (such as the status of routes and points) is available in the RBC. Thus, an image of the supervised area exists in the RBC. With this information, the RBC can calculate a movement authority for every ETCS controlled train in its area.

The ETCS on-board processes the information received from the RBC and continuously updates the train speed and position information, comparing it with respect to the speed and distance supervision limits conveyed by the movement authority. All the information to enable driving within these supervision limits is presented to the driver on the DMI.

In contrast to level 1, the ETCS level 2 on-board also sends information to the RBC, for example a movement authority request or a position report. Based on the position report, the train position on the RBC route map is updated. There is also a monitoring function which ensures that in the event of a radio communication failure, the train triggers a defined safety reaction after a configurable time period has elapsed.

Level 2 supports three methods of train separation, depending on the train detection methods that are used.

Train Separation Method	Description	On-board Train Integrity
Fixed Block	Safe train separation is based on trackside train detection sections, e.g. track circuits, axle counters.	Not required. Protection against a split train is provided by virtue of the trackside train detection sections.
Virtual Block (previously known as Hybrid level 3)	Safe train separation is based on a limited number of longer trackside train detection sections that are sub-divided into smaller virtual sub-sections.	Optional To make use of the optimised train spacing that is available through the virtual sub-sections, train integrity is required, reported by the ETCS on-board to the RBC as part of the position report. Safe separation of trains that are not fitted with train integrity equipment is ensured using the longer trackside train detection sections.
Moving Block (previously known as level 3)	Safe train separation is based on the position reports of trains alone. This enables optimal moving block spacing between trains.	Required Train integrity is required for all trains, reported by the ETCS on-board to the RBC as part of the position report.

Table 7-2: ETCS level 2 train separation methods

Where train integrity is required, it can be achieved by several different methods. For fixed formation or multiple unit passenger trains it is relatively easy to allocate train wires (wires that run through the length of the train) for the ETCS on-board to monitor. For locomotives coupled to individual coaches or wagons it may be impractical to use train wires, and for this reason there are train integrity management systems that use 'head of train' and 'tail of train' units that are fitted with sensors and communicate with one another (over radio). By comparing speed, acceleration and other parameters these units can establish whether the train is intact or not.

The ETCS on-board uses the train integrity to calculate the 'confirmed train length' by subtracting the 'estimated front end of the train' from the 'minimum safe rear end' of the train when integrity was last confirmed. The confirmed train length is then sent to the RBC as part of the position report.

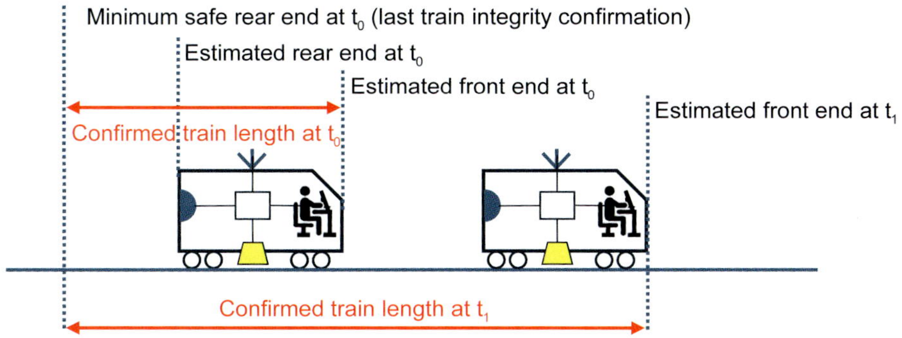

Figure 7-6: Confirmed train length calculation

ETCS also allows for train integrity to be confirmed by the driver via the DMI but only when the train is at a standstill. This means that the confirmed train length can 'grow' significantly between stops, making this method unsuitable for a true moving block application. It could however be useful to avoid the need for trackside train detection on a secondary line with infrequent trains.

The key elements and characteristics of ETCS level 2 are:

Trackside equipment:

- Radio Block Centre
- Radio network for bi-directional track-train communication
- Balises, provided mainly for location referencing

Main ETCS trackside functions:

- Know the ETCS identity of each train that is equipped with, and running under, ETCS within an RBC area
- Monitor the location of each ETCS controlled train within an RBC area
- Determine movement authorities for each train individually, according to the status of the underlying signalling system
- Transmit movement authorities and track description data (gradient, speed limits, etc.) to each train individually
- Transmit other information to the train (e.g. text message to the driver or emergency stop command)
- Handing over of train control between different RBCs at the RBC-RBC borders

Main ETCS on-board functions:

- Read balises and send the train position (relative to the detected balises) to the RBC; this also includes train integrity status if available
- Receive a movement authority and the track description via radio
- Select the most restrictive speed profile
- Calculate a dynamic speed profile, taking into account the distance to the end of the movement authority, the track description data and the train running/braking characteristics (which are known to the ETCS on-board)
- Compare the actual train speed with the permitted speed (from the speed profile), and command a brake application if necessary
- Display cab signalling information to the driver via the DMI

7.5 Level NTC (National Train Control)

'National Train Control' (NTC) refers to train control (protection) systems other than ETCS. These are referred to as 'Class B systems' in the TSI CCS and are generally older systems that were developed for a national railway operator by a single supplier. The ultimate objective is to achieve full interoperability by replacing these NTC systems with ETCS, but this process may take several decades and is dependent on each nation's ETCS deployment plans.

The degree of supervision and safety for a NTC system can vary from an intermittent system sending only a stop/no stop message, to a continuous speed supervision system that sends a large number of messages from the national system's trackside equipment to the on-board equipment. The means of communication between track and train varies from simple trackside magnets and coded track circuits to beacons (similar to ETCS balises), inductive loops along the track, and radio systems such as Wi-Fi.

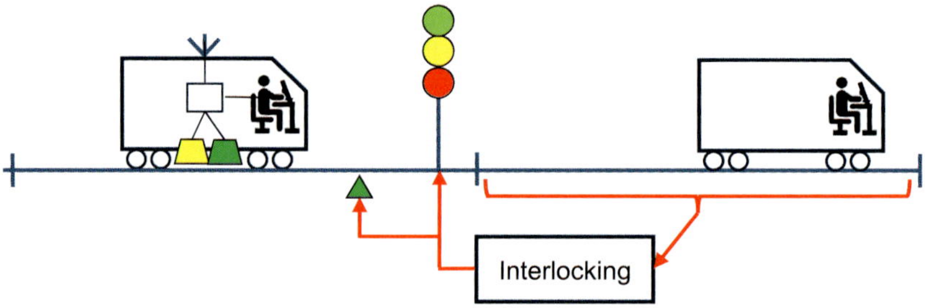

Figure 7-7: ETCS level NTC

Level NTC allows trains that are equipped with ETCS to run on lines equipped with a national system. When the ETCS on-board enters level NTC it hands over control and safety responsibility to on-board equipment for the national system, including the transmission equipment that communicates with the trackside equipment for the national system (green symbols in figure 7-7). There are various options for implementing the on-board hardware for the national system, including a 'Specific Transmission Module' (STM) which gives the national system access to some of the resources of the ETCS on-board, e.g. train interface, DMI, odometry. Refer to chapter 20 for a more detailed explanation of these options.

The ETCS on-board system can transition between level NTC and ETCS levels 0, 1, 2 in response to balise telegrams, meaning that a train can travel seamlessly along a line with a mix of trackside (ETCS and NTC) systems. Level NTC is activated by a level transition order from the trackside or can be manually selected by the driver.

The ETCS on-board DMI is normally used to give the driver information received from the national system, but specifically what information, and how it is presented, is defined by the national system specification. The ETCS DMI specification (ERA_ERTMS_15560) includes requirements for the display of NTC related information. The national system can then make use of the indications and buttons defined for the ETCS on-board and can also define additional or alternative indications and buttons to be used by the driver when running in Level NTC.

On some implementations the national system will have its own separate DMI. The driver must then make use of the relevant DMI depending on current level of the ETCS on-board. For example, level transition to another level must be acknowledged by the driver using the ETCS

DMI. At the same time, the national system could require a driver response on the national system's separate DMI.

National System status

The Specific Transmission Module FFFIS (SUBSET-035) defines a number of national system statuses that are used in the level transitions to and from level NTC, and these may also be used when another type of interface to the national system is implemented:

- Cold standby
- Hot standby
- Data available
- Failure

Before a level transition to a NTC area is announced, the national system is in 'cold standby' status, with its trackside communication switched off. When a level transition announcement has been received by the ETCS on-board system, the national system is ordered by the ETCS on-board to change state to 'hot standby'. When in this state, the national system will initiate communication with the national trackside equipment and store the appropriate trackside information to be used after the level transition border.

When the level transition border is passed, the national system is ordered by the ETCS on-board to enter 'data available' state. In this state the national system communicates with its trackside and is responsible for the supervision of safe train movement.

The national system's status is controlled by the ETCS on-board system. The only exception is when a failure is detected by the national system's on-board equipment. In such cases, the national system will change its status to 'failure'.

ETCS functions in level NTC

In level NTC the ETCS on-board operating in SN (National System) mode still performs various functions in the background and has the responsibility of taking safe action in case the communication with the national system's on-board equipment is lost.

The ETCS on-board processes balise telegrams relating to the following functions in preparation for a transition out of level NTC:

- National values
- Level transition orders
- Communication session management
- Train position

The ETCS on-board is responsible for keeping track of train position by reading balises and using its odometry equipment. ETCS train position will also be reported to an RBC when running in SN mode if a communication session has been established on the approach to a level 2 transition border. Depending on how the information is used in the RBC, it might be important to install some balise groups in the national system area in order to avoid the confidence interval growing to such a large value that the ETCS train position information cannot be used.

When a level transition out of level NTC has been announced, information related to the new level will be stored for use after the level transition border.

Start of Mission in level NTC

In level NTC the standard ETCS Start of Mission procedure (see chapter 9) is performed but, depending on the national system, the train data entry procedure might include some additional data required by the national system's on-board equipment. If several national systems are installed on-board, the train data procedure will, once the ETCS train data has been entered, request the driver to enter all the additional national system data for all of the national systems that require data entry. The train data entry procedure is not finalised until all the installed national systems have received the required additional data. This is done to enable seamless level transitioning while the train is being driven.

7.6 Mixed ETCS levels

It is possible to superimpose several ETCS levels in parallel on the same track. Take, for example, a station which is shared by trains arriving on level 1 and level 2 lines, or parallel operation of a national system with ETCS. Another reason could be to allow differently equipped trains to run on the same line.

Superimposing more than one level on a line could also be implemented to have a backup if one of the levels fails, whether trackside or on-board. A train could continue after the driver selects another of the levels supported by the trackside (included in the 'priority list' sent from balise group or RBC together with the level transition order). Note that the priority list is contained within the level transition order data packet, described further below. There are no restrictions on the levels that could be superimposed in the ETCS standard.

On lines equipped for mixed levels, trackside equipment must be installed for all the superimposed levels. ETCS level 2 overlay is a good example of this, whereby the trackside will include radio coverage and balises (for ETCS level 2) and also NTC equipment (for ETCS level NTC) associated with the remaining lineside signals.

The ETCS on-board system in trains running on a mixed level line selects the appropriate level according to the equipment installed on-board, and according to the priority list sent from the balise groups or the RBC. In the ETCS level 2 overlay example, the priority list will normally have level 2 as the highest (preferred) priority, such that an ETCS equipped train will transition to level 2 in the overlay territory.

The priority list of trackside supported levels will be stored by the ETCS on-board until the system is powered off. Note that the cold movement detection function described later in this book is designed to retain the previously applicable level and the priority list of supported levels. This means that, even after powering off, when the ETCS is powered on this information can be validated and used, provided that the train has not moved.

There are two ways to inform the ETCS on-board system which levels are supported/allowed trackside:

- using a 'level transition order'
- using a 'conditional level transition order'

In both cases, all the trackside supported levels are listed with an assigned priority, and new level transition information received from trackside replaces previously received information.

In the case of a level transition order, the ETCS on-board will select the highest priority level that it can support when receiving a level transition order. In the case of a conditional level transition order, the ETCS on-board will keep to the current level if this level is included in the priority list received from the trackside, regardless of the assigned priority. If the ETCS on-board is running at a level that is not included in the priority list received from the trackside, the ETCS on-board will immediately switch to the highest priority level that it can support.

Once the train is at standstill, the driver can request to change level on the DMI. When requested, a list of possible levels is shown to the driver. The list is derived from the priority list of levels supported by the trackside and the levels supported by the on-board. There are no restrictions on the driver changing the level, other than the train must be at standstill (a technical requirement), and that the driver should seek permission from the control centre operator (an operational requirement).

When changing levels from level 0 or level NTC to level 1 or level 2 by driver selection, the ETCS on-board will not initially have a valid movement authority, and this will lead to a 'Trip' condition (level transition with no valid MA). The Trip (TR) mode calls for an irrevocable application of the emergency brake. The handling of this situation is not specified in the standard and is therefore implemented differently by different suppliers. One possibility for avoiding the Trip condition could be to instruct the driver to press the 'override end of authority' button on the DMI before selection of level 1 or level 2.

The last used level (previously applicable level) will be stored by the ETCS on-board when the power is turned off. When the ETCS on-board system is powered up (and cold movement detection has validated the previously applicable level), the last used level will be proposed to the driver during the start of mission.

7.7 Level transitions

Trains running across areas with differently equipped trackside infrastructure must be informed about the approaching change of level, known as a 'level transition'. Announcement of a level transition is required in order for the ETCS on-board to accept and store information prior to the level transition border. The announcement of a new level can be sent by balise groups or from the RBC, and it includes the distance to the level transition border. The level transition border itself is marked by a balise group that gives the immediate level transition order (the order is in fact another level transition announcement with distance set to zero).

When an immediate level transition order is received, the ETCS on-board performs the level transition regardless of the current operational mode. This enables all ETCS on-board units on a train to be synchronised in terms of level. A train may have multiple ETCS on-board units, one of which might be in FS (Full Supervision) mode, with the others in SL (Sleeping) mode.

If the balise group giving the immediate level transition order is missed, due to a faulty or missing balise group, the level transition will still be performed when the distance to the level transition border given in the announcement of the level transition has been passed by the train.

The announced level will be presented to the driver. Some level transitions, such as transition to level 0 or level NTC, will require an acknowledgement from the driver to make them aware of their increased responsibility because of the reduced level of train protection and supervision afforded by the on-board system. A level transition to a similar level of supervision will not require the driver's acknowledgment, e.g. transition from level 2 to level 1. The level transition must be acknowledged by the driver, when required, before the level transition border to avoid a brake intervention by the ETCS on-board.

When entering an ETCS area (level 1 or 2), the ETCS on-board must have already received a movement authority (MA) and track description data covering the full length of the MA. The on-board system will change to TR (Trip) mode and apply the emergency brake at the level transition border if no MA and/or track description data has been received and stored before the border. A level transition order to level 1 can be given in the same balise group as the MA and track description data, but in level 2 the information must be sent from the RBC after the level transition announcement but before reaching the border, to avoid a Trip. New information can be sent to the train if the conditions change while the train is still in the announcement area.

To make sure the ETCS on-board and the trackside have the same interpretation of the conditions after the level transition border, rules have been established regarding which and how many messages can be accepted for storage during the announcement distance. This is to avoid a situation where, for example, the RBC sends a lot of information to the ETCS on-board without knowledge of any limits in the ETCS on-board storage.

The ETCS on-board must store no more than three sets of information from three filtered messages (i.e. accepted for storage). Additional information will replace the oldest of the stored sets of information (on the principle of “first in, first out”). By setting this limit of three, the RBC could for example send three emergency stop messages to make sure that any previously sent MA is removed.

The same general approach applies during RBC-RBC handover, which is in effect a level 2 to level 2 transition when a train passes between the control areas of two different RBCs. The accepting RBC (the RBC controlling the area which the train is about to enter) can send information to the ETCS on-board before the train has passed the border and that information will be stored on-board following the same principles as for a level transition.

Refer to chapter 9 for further explanation of some typical level transition scenarios and chapter 22 for more detail of the RBC-RBC handover.

References

SUBSET-026-02: System Requirements Specification, Chapter 2 Basic System Description

SUBSET-035: Specific Transmission Module FFFIS

ERA_ERTMS_15560: ETCS Driver Machine Interface

Chapter 8: Modes and their Functionality

8.1 Introduction to ETCS modes

This chapter explains the various ETCS on-board equipment modes. A brief explanation of what is meant by an ETCS mode is provided, all currently permissible ETCS modes are listed, and their usage is described.

The term 'ETCS modes' refers to the different operating modes in which the ETCS on-board equipment can function. A mode is activated either by the ETCS trackside equipment automatically, or by the driver manually, depending upon the operating conditions.

Each mode enables the train to be driven in accordance with a specific set of operational and technical requirements and constraints, providing operators with the ability to perform a wide range of train movements such as normal running, shunting, joining and splitting of trains, and movement under system failure conditions. Although some modes are considered mandatory (they cannot be disabled), other modes can be made available or disabled depending on the operational requirements.

The following modes are provided for in the version 4.0.0 of the ETCS System Requirements Specification SUBSET-026. It is interesting to note that since the first edition of this book was published four new modes have been added, as indicated.

- Full Supervision (FS)
- Automatic Driving (AD) Added in Baseline 4
- Limited Supervision (LS) Added in Baseline 3
- On Sight (OS)
- Staff Responsible (SR)
- Shunting (SH)
- Supervised Manoeuvre (SM) Added in Baseline 4
- Passive Shunting (PS) Added in Baseline 3
- Unfitted (UN)
- Sleeping (SL)
- Stand By (SB)
- Trip (TR)
- Post Trip (PT)
- System Failure (SF)
- Isolation (IS)
- No Power (NP)
- Non Leading (NL)
- National System (SN)
- Reversing (RV)

Later in this chapter two examples of transitions between modes are provided. The examples describe the main conditions under which a mode transition is performed from Staff Responsible (SR) mode to Full Supervision (FS) mode and back again, in conjunction with the associated operating procedure.

The formal description of each mode is found in the ETCS System Requirement Specification SUBSET-026, Chapter 4 “Modes and Transitions”, which covers:

- Definition of the modes
- Listing of vehicle functions active in the modes
- Conditions for transition from one mode to another

- Information and operator control actions available to the driver in a mode
- Trackside information accepted by the on-board unit
- Handling of information already saved for the mode transition

8.2 Description of each mode

Full Supervision (FS)

Full Supervision (FS) mode, along with Automatic Driving (AD) mode (see below), is the highest level of supervision and protection for a train.

FS mode cannot be selected by the driver. The transition from another mode to FS mode is performed automatically if the ETCS on-board equipment has received and evaluated a full movement authority (MA) together with the associated profile data (static speed profile and gradient profile), and has the necessary train data. The ETCS on-board supervises the train movement against the dynamic speed profile, and displays information such as actual speed, target speed, permitted speed, release speed, gradients and distances on the driver machine interface (DMI). Responsibility for train protection lies fully with the ETCS system, although the driver remains in control of the train movement provided that the dynamic speed profile is respected at all times.

Note that the DMI can also display driver advisory system (DAS) information when in FS mode. To achieve this, the train must be equipped an Automatic Train Operation (ATO) on-board system configured to provide the DAS information. The DAS information displayed on the DMI can include dwell time, skip stop advice, stopping point distance, name and estimated arrival time, stopping accuracy, door information, target advice speed or coasting advice.

This mode can be used in ETCS level 1 and level 2.

Automatic Driving (AD)

Automatic Driving (AD) mode requires the train to be equipped with an ATO on-board system which is interfaced to the ETCS on-board (see chapter 13 for more details of ETCS with ATO).

The preconditions for entering AD mode (all of which must be fulfilled) are that:

- The ATO on-board has received a valid journey profile from the ATO trackside.
- The ETCS on-board is in Full Supervision (FS) mode immediately prior to the transition to AD mode.
- The driver has selected 'ATO engage' on the DMI.

In AD mode responsibility for train protection lies fully with the ETCS system, supervising the ATO system which is driving the train. The driver can at any time take back control of the train by selecting 'ATO disengage' on the DMI, or by applying the brake. Either action causes the ETCS on-board to transition from AD mode back to FS mode.

For passenger trains the ATO is automatically disengaged at each stopping point (station). After the station activities are complete and the driver has closed the doors (and other conditions as defined in the ATO standard are met), the 'ATO engage' button on the DMI will be enabled, ready for the driver to select it again. Thus the ETCS on-board mode will change from AD mode (when the train is driving towards the stopping point) to FS mode (when the train has stopped and the ATO is automatically disengaged) and then back to AD mode when the train is about to depart (after the driver has selected 'ATO engage').

In AD mode the DMI conveys some information differently from FS mode, to avoid the driver from feeling the need to intervene. For example, the audible alerts 'Sinfo-S1' (Overspeed) and 'Sinfo-S2' (Warning) are suppressed and the target indications on the planning area are displayed in white, instead of yellow. There is an additional button on the DMI for ATO engage/disengage/status, and additional ATO information is provided relating to stopping points (stopping accuracy, dwell time, door information) and running sections (next stopping point, estimated arrival time, skip stop advice).

This mode can be used in ETCS level 1 and level 2.

Limited Supervision (LS)

Limited Supervision (LS) mode enables the ETCS on-board to be operated in areas with signals but only partially equipped with ETCS trackside equipment. For example, in level 1 LS, not all signals are required to be equipped with balises. As well as being a lower-cost ETCS solution in its own right, partial trackside fitment with the use of the LS mode can provide a migration path to full ETCS deployment.

The transition to LS mode is performed automatically when commanded by trackside, e.g. balise telegram or radio message, and all necessary conditions are fulfilled (these are determined by the infrastructure manager). As with FS mode, the driver cannot select LS mode. The ETCS on-board supervises the train movement against the dynamic speed profile, but the DMI only displays actual speed and release speed and, if ordered by the ETCS trackside, the 'lowest supervised speed within the movement authority' (LSSMA). Responsibility for train protection lies with both the ETCS system (to the extent permitted by the information supplied by the trackside ETCS) and the driver.

This mode can be used in ETCS level 1 and level 2.

On Sight (OS)

On Sight (OS) mode enables the train to enter a track section that could already be occupied by another train or obstructed by any kind of obstacle. OS mode supports planned operational scenarios such as train joining, as well as permitting train movements under failure conditions, such as an axle counter that is wrongly causing the track section to show occupied.

The transition to OS mode can only take place if commanded by the ETCS trackside (it cannot be selected by the driver). The trigger for entering OS mode originates from the traffic control system in the form of a route request. The route request is acted upon by the interlocking and the route status is then sent to the ETCS trackside (RBC or balise), resulting in an OS mode profile command to the ETCS on-board. Note that the specific conditions for allowing a route to be set under the conditions outlined above (track section occupied or obstructed) will depend on the signalling principles that the infrastructure manager applies to this type of train movement.

The driver must acknowledge a change to OS mode when it is displayed on the DMI, because this mode requires a high degree of driver responsibility. If not acknowledged within five seconds, the ETCS on-board will trigger the service brake to be applied. OS mode gives the driver the responsibility for checking the track for any obstacles, although the ETCS on-board continues to supervise the train speed. The DMI displays information such as actual speed, target speed, permitted speed, release speed, gradients and distances on the DMI. There are differences in how this information is displayed, compared with other modes, in order to make it very obvious (in conjunction with the OS mode symbol) that responsibility for safe movement of the train is shared by the driver and the ETCS on-board.

This mode can be used in ETCS level 1 and level 2.

Staff Responsible (SR)

Staff Responsible (SR) mode allows a train to be moved under the driver's own responsibility in an ETCS equipped area. Typically, SR mode is used in failure scenarios, for example if the ETCS on-board has no train position information, or to pass a lineside signal at danger.

In SR mode the ETCS on-board supervises the train movement against a ceiling speed and a permitted distance to move. Balise groups located beyond the permitted distance command the train to 'Stop if in SR', and balises also command the train to stop if a reverse movement is detected (i.e. if the train passes over a balise in the wrong direction). The ETCS on-board can also supervise the train movement against a mode profile or list of balise groups (if sent by the RBC). The maximum allowable SR ceiling speed and SR distance are set by national values (V_NVSTFF and D_NVSTFF), although they can also be entered manually by the driver. The permitted movement distance in SR in ETCS level 2 can also be set via a message from the RBC. Responsibility for train protection lies primarily with the driver.

This mode can be used in ETCS level 1 and level 2.

Shunting (SH)

Shunting (SH) mode allows a train to be moved under the driver's own responsibility within a shunting area. Reversing is permitted in this mode. The ETCS on-board supervises the train's movements against a ceiling speed. The maximum allowable SH ceiling speed is set by the national value V_NVSHUNT. Balise groups can be set to define the limits of the movements by giving the order 'Stop if in SH Mode' if the train passes over them. The ETCS on-board can also supervise the movements against a mode profile or a list of balise groups (if sent by the RBC).

The ETCS on-board equipment can be switched to SH mode by a command from the ETCS trackside equipment, or by the driver. To ensure that the driver is aware of the increase in driver responsibility, the driver is required to acknowledge a change to SH mode that has been ordered by the trackside. If no acknowledgement is received within five seconds, the ETCS on-board will trigger the service brake to be applied. No acknowledgement is required if SH mode was requested by the driver, but if in level 2 the RBC might refuse the request if the conditions for entering SH mode are not met.

Responsibility for shunting movements and for compliance with the limits of the shunting area lies with the driver, noting of course that this may also be supervised through the measures outlined above.

This mode can be used in ETCS level 0, level 1, level 2, and level NTC.

Supervised Manoeuvre (SM)

Supervised Manoeuvre (SM) mode enables enhanced supervision of shunting movements (compared with SH mode), the level of supervision being similar to that of FS mode but allowing for the possibility that the driving cab may not be at the front of the vehicles making the shunt movement (referred to as the 'consist' below).

SM mode is requested by the driver selecting 'Initiate SM' on the DMI main menu when the train is at a standstill. Provided that all the conditions for SM mode are met, the RBC sends a 'SM Authorisation' message to the ETCS on-board. This message contains a movement authority, gradient profile, static speed profile and SM mode speed limit.

One of the most important conditions for entering SM mode is that the RBC receives the 'safe consist length' from the ETCS on-board. This describes the train length in relation to the position of the driving cab within in the overall train consist.

In SM Mode, the DMI displays a 'SM authorised direction indicator' (an arrow) within the circular speed indicator that tells the driver if the SM movement authority is forwards (arrow up) or backwards (arrow down) in relation to the normal driving direction as viewed from the cab.

This mode can only be used in ETCS level 2.

Passive Shunting (PS)

Passive Shunting (PS) mode is restricted to a locomotive fitted with one ETCS on-board and two driving cabs, or two separate locomotives that are mechanically but not electrically coupled, each with its own ETCS on-board. It is designed to make the process of shunting more efficient by removing the need for the 'start of mission' procedure to be applied every time the driver changes driving cab.

PS mode requires a special passive shunting input from the vehicle. This input has two states; 'passive shunting permitted' and 'passive shunting not permitted'. The passive shunting permitted state confirms that the consist has been immobilised by the braking system and that the brakes can be released from another cab after activation. This is important because in PS mode the ETCS on-board has no responsibility for train protection. The transition to PS mode is only allowed if the ETCS on-board knows that the train has been safely immobilised.

PS mode is primed when the driver selects 'Maintain Shunting' (step 1 in figure 8-1). When the desk is closed, the ETCS on-board checks if the passive shunting input is set to 'passive shunting permitted' and then transitions to PS mode (step 2). The selection of PS mode causes the on-board to retain information such as balise lists that were previously applicable. PS mode transitions back to SH mode when the driver re-activates the cab (opens the desk) provided that the on-board has not picked up a 'stop shunting on desk opening' message from the trackside (step 3).

In PS there is no train protection provided by the ETCS and the notion of driver responsibility is not relevant (because the desk is closed).

This mode can be used in ETCS level 0, level 1, level 2, and level NTC.

1. Driver selects 'Maintain Shunting'.

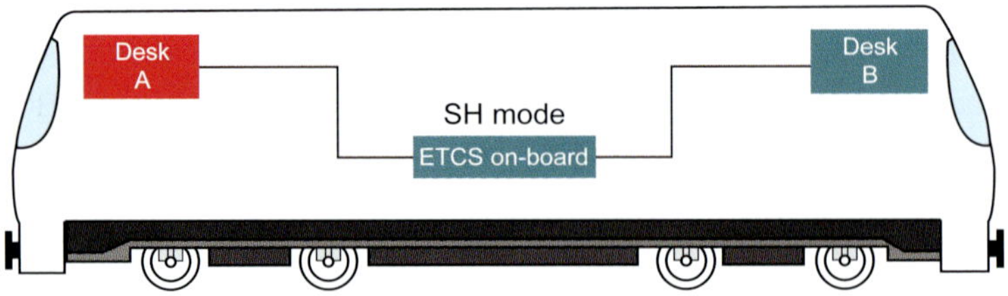

2. Driver closes Desk A and the ETCS transitions to PS mode provided that the 'passive shunting permitted' input is set.

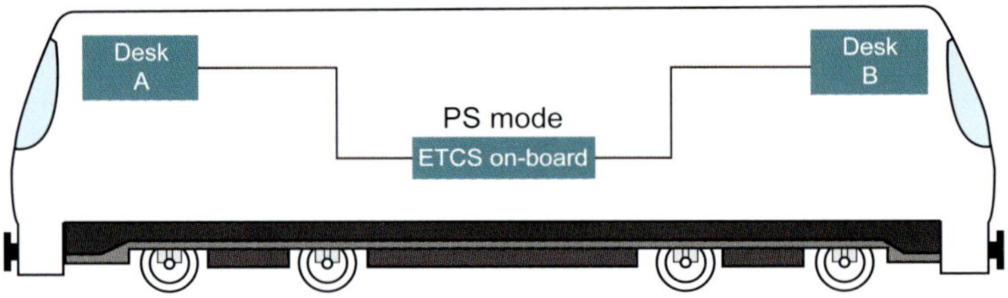

3. Driver opens Desk B and the ETCS immediately transitions to SH mode (no need for train data entry).

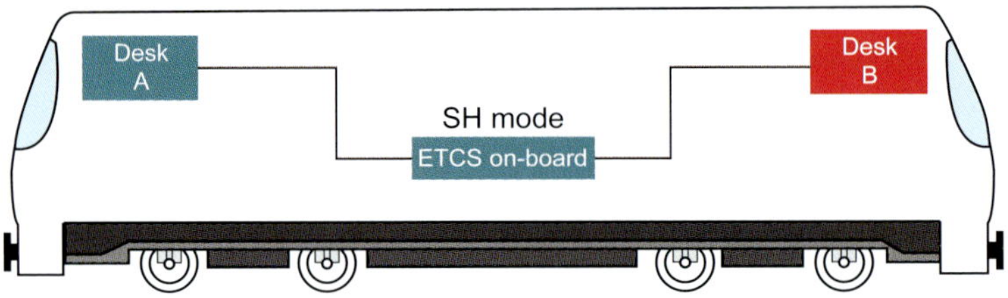

Figure 8-1: SH and PS modes on a locomotive with one ETCS on-board and two cabs

Unfitted (UN)

Unfitted (UN) mode enables a train movement to be made on track that has no trackside train protection equipment (neither ETCS nor a national train control system). This might apply in a depot, or where the trackside equipment is fitted but not in service. The ETCS on-board supervises train movements against the UN ceiling speed (as defined by national value V_NVUNFIT).

UN mode is entered when the driver selects level 0, or when level 0 is commanded by a 'level transition order', in which case it must be acknowledged by the driver, and the ETCS on-board will apply the service brake if the driver does not do so. In UN mode responsibility for train protection lies with the driver.

This mode can be used in ETCS level 0 only.

Sleeping (SL)

Sleeping (SL) mode is the default mode for the ETCS on-board when the associated driving cab is closed (inactive) and the train ('consist') of which that cab is a part is electrically connected via the train interface to another vehicle in which the driving cab is open (active). This mode is provided so that the ETCS on-board associated with an inactive driving cab maintains its knowledge of the train's position. In this way, when the cab is made active again (and after the driver has completed the Start of Mission procedure), the ETCS on-board can start directly in FS or OS mode. This facilitates efficient turnaround operation.

SL mode is entered automatically when the ETCS on-board is in SB mode (or PS mode) and subsequently detects the 'sleeping' input signal via the train interface. The train interface ensures that when one cab is active, every ETCS on-board in another cab in the consist will receive this input. This input is considered to be safety critical because in SL mode there is no standstill supervision (standstill supervision is applied only when the ETCS is in SB mode).

When an ETCS on-board and the associated driving cab is in SL mode, the cab is not being used for driving the train. Therefore, there is no train protection provided by the ETCS and the notion of driver responsibility is not relevant.

This mode can be used in ETCS level 0, level 1, level 2, and level NTC.

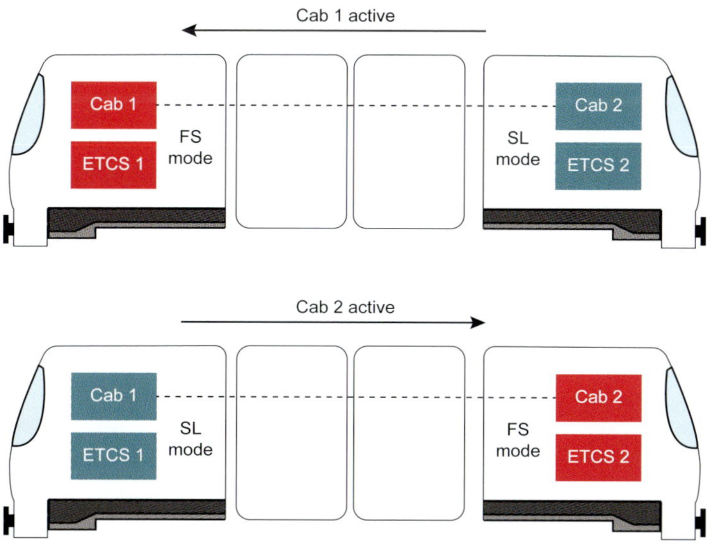

Figure 8-2: SL mode with an ETCS on-board at each end of the train

Stand By (SB)

Stand By (SB) mode is a default mode that cannot be selected by the driver.

When the ETCS on-board is powered up it initially enters SB mode, irrespective of the driving cab ETCS status (open/closed). The ETCS on-board provides standstill supervision in SB mode so that if the train moves beyond a specified distance (as defined by national value D_NVROLL), the brake will be applied.

Self-tests and peripheral tests are performed in SB mode. Where required by national standards this will include a brake text. Whilst in SB mode, train data can be entered by the driver as part of the Start of Mission procedure. The results of the tests and the data entry are shown to the driver on the DMI.

This mode can be used in ETCS level 0, level 1, level 2, and level NTC.

Trip (TR)

When certain safety critical events occur (e.g. train passing the end of its movement authority) the ETCS on-board enters Trip (TR) mode. This initiates the emergency brakes, and the ETCS on-board monitors the train until it comes to a standstill. The driver is required to acknowledge the Trip, and once acknowledgement has been received, the ETCS on-board equipment switches to Post Trip (PT) mode (if in level 1 or level 2), or to UN/SN/SH mode (if in level 0 or level NTC). Note that the brakes cannot be released whilst in TR mode.

TR mode is different to an emergency brake application that may occur in other circumstances. For example, if the train exceeds the permitted speed an emergency brake application can be commanded by the ETCS, but the brake application can be revoked when the train is no longer exceeding the permitted speed (provided that the national value (Q_NVEMRRLS, Emergency Brake Release) is set). By contrast, a Trip forces the train to a standstill and requires driver acknowledgement in order to exit TR mode and return to normal operations.

This mode can be used in ETCS level 0, level 1, level 2, and level NTC.

Post Trip (PT)

The ETCS on-board enter this mode after acknowledgement of TR mode by the driver in level 1 or level 2. In Post Trip (PT) mode the train cannot move forward, although a reversing movement within a maximum permitted distance (as defined in the national value D_NVPOTRP) is possible.

After entering PT mode, the procedure for re-starting the forward train movement depends on the ETCS level. In level 2, the RBC must firstly acknowledge the exit from TR mode before the driver can make any selection on the DMI. Typically, the driver will select ‘Start’ which will send a MA request to the RBC. Depending on the response from the RBC, the ETCS will transition directly to FS, or offer either OS or SR mode for driver acknowledgement. In level 1, the driver can select the next action on the DMI as soon as the ETCS on-board has entered PT mode. Typically, the driver will select ‘Start’ in which case the ETCS will offer SR mode for driver acknowledgement.

In PT mode the ETCS on-board is responsible for supervising any reversing movement, including ensuring that the movement does not exceed the maximum permitted distance.

This mode can be used in ETCS level 1 and level 2.

System Failure (SF)

System Failure (SF) mode is entered by the ETCS on-board equipment if a safety related error occurs. The emergency brakes are activated and remain permanently applied. To continue operation the driver must either get permission to reset (power off and then power on again) or isolate the ETCS on-board.

This mode can be used in ETCS level 0, level 1, level 2 and level NTC.

Isolation (IS)

Isolation (IS) mode is entered when the driver operates an isolation device, typically a key switch in the cab. In IS mode the ETCS on-board is physically isolated from the brakes and cannot receive information from either the train or from the trackside equipment.

The train is the exclusive responsibility of the driver in this mode. In order to leave IS mode, a special operating procedure must be applied, to ensure that the ETCS on-board is only returned to service when it has been proven to be safe for operation.

Note that in many applications the DMI will continue to display the train speed in IS mode, using speed data from either the ETCS on-board or from an alternative speed source. The choice of how to display the speed when in IS mode is project specific and is not governed by the ETCS standards.

This mode can be used in ETCS level 0, level 1, level 2, and level NTC.

No Power (NP)

In No Power (NP) mode the main ETCS on-board equipment is not powered and the ETCS on-board permanently commands the emergency brake.

The ETCS on-board is designed such that the cold movement detector remains powered via a separate power supply (typically directly from the train battery). On entry to NP mode the train position and ETCS level are both stored but tagged as 'invalid'. On exiting NP mode and providing that no movement has occurred (as indicated by the cold movement detector), the train position and ETCS level are returned to the 'valid' state.

This mode can be used in ETCS level 0, level 1, level 2, and level NTC.

Non Leading (NL)

Non Leading (NL) mode is designed to manage the ETCS on-board on a 'slave' locomotive coupled behind another locomotive in tandem operation with drivers in the cabs of both locomotives, each controlling traction independently.

NL mode can be selected by the driver when the on-board detects the 'non leading permitted' input via the train interface. This confirms that the slave locomotive is either isolated from the train brake completely, or if connected, that the driver can only apply and not release the brakes.

In NL mode there is no train protection provided by the ETCS, but the ETCS on-board maintains its position and, in the case of level 2, reports the position to the RBC. In NL mode the ETCS on-board displays the speed on the DMI and displays track condition information for the driver to respond if relevant (e.g. lower pantograph).

This mode can be used in ETCS level 0, level 1, level 2, and level NTC.

National System (SN)

National System (SN) mode is entered when the ETCS on-board has passed responsibility for train protection to a national train control (NTC) system. The national train control system can be implemented either as a specific transmission module (STM) using a standardised interface to the ETCS on-board or using an alternative solution (see examples in chapter 20).

From the driver's perspective the DMI will display the identification of the national train control system (NID_NTC), and the SN mode symbol. The right-hand side of the DMI may display buttons and indications for the national system, depending on the design and implementation. The degree of responsibility of the driver for the train's movement depends on the functionality of the national train control system in use.

This mode can be used in ETCS level NTC only.

Reversing (RV)

In Reversing (RV) mode, the driver is permitted to change the direction of movement of the train without changing cabs. RV mode can be entered only in areas where permission has been transmitted to the ETCS on-board by the ETCS trackside equipment.

When the front end of the train is stationary inside the reversing permitted area, the DMI indicates to the driver that reversing is possible. When the ETCS on-board subsequently detects that the driver intends to reverse (e.g. direction controller has been moved to the reverse position), the driver is required to acknowledge RV mode on the DMI in order to enter RV mode and start the reverse movement.

When in RV mode, the ETCS on-board supervises the maximum permitted speed and maximum distance that the train is allowed to move in the reverse direction. These supervision parameters are transmitted to the on-board in advance by the ETCS trackside. If this maximum distance is exceeded, the ETCS on-board activates the emergency brake. The ETCS on-board also applies unauthorised movement direction protection to prevent the train from moving in the forward direction.

In RV mode the DMI displays train speed, permitted speed and the remaining distance to run.

This mode can be used in ETCS level 1 and level 2.

8.3 Transitions between modes

Various operational and technical conditions must be fulfilled in order for a transition from one mode to another mode to occur. A complete overview of all the conditions is provided in the System Requirement Specification SUBSET-026, Chapter 4 Modes and Transitions in the form of a 'transition table' (figure 8-3).

NP	<29 -p2-	<29 -p2-	<29 -p2-	<29 -p2-	<29 -p2-	<29 -p2-	<29 -p2-	<29 -p2-	<29 -p2-	<29 -p2-	<29 -p2-	<29 -p2-	<29 -p2-	<29 -p2-	<29 -p2-		<29 -p2-	<29 -p2-
4> -p2-	**SB**	<22 -p4-	<19, 27, 30 -p5-	<28, 82 -p4-	<28,83 -p5-	<28,83 -p5-	<28,83 -p5-	<28,83 -p5-	<28,83 -p5-	<2, 3 -p3-	<28, 47 -p3-	<28,83 -p6-		<28,83 -p4			<28,83 -p6-	<28 -p4-
		PS	<26 -p5-															
	5, 6, 50> -p7-	23> -p4	**SH**	<6 -p6-	<5,6, 50,51 -p6-	<5,6, 50,51 -p6	<5,6, 50,51 -p6-	<5,6, 51 -p6-	<5,6 50,51 -p6-			<5,61 -p7-	<68 -p4	<5,6, 50,78 -p5-			<5,61 -p7	
	81> -p8-			**SM**	<81 -p7-	<81 -p7-	<81 -p7-	<81 -p7-	<81 -p7-					<81 -p6-				
	10> -p7-				**FS**	<9,24, 33,48, 53 -p7-	<76 -p6-	<31,32 -p6-	<75 -p6-			<25 -p7-		<31 -p5-			<25 -p7-	
					80> -p7-	**AD**												
	70> -p7-				70,72> -p6-	70,72> -p6-	**LS**	<72 -p6-	<70,74 -p6-			<71 -p7-		<70 -p5-			<71 -p7-	
	8,37> -p7-				37> -p6-	37> -p6-	37> -p6-	**SR**	<37 -p6-			<44,45 -p4-		<8,37 -p5-			<44,45 -p4-	
	15> -p7-				15,40> -p6-	15,40> -p6-	15,73> -p6-	40> -p6-	**OS**			<34 -p7-		<15 -p5-			<34 -p7-	
	14> -p5-	14> -p4								**SL**								
	46> -p6-		46> -p5-	46> -p6-	46> -p6-	46> -p6-	46> -p6-	46> -p6-	46> -p6-		**NL**							
	60> -p7-				21> -p6-	21> -p6-	21> -p6-	21> -p6-	21> -p6-			**UN**	<62 -p4-	<77 -p5-			<21 -p7-	
	20> -p4-		49,52, 65> -p4-	16,17, 20,41, 65,66, 69> -p5-	11,12, 16,17, 18,20, 41,65, 66,69> -p4-	11,12, 16,17, 18,20, 41,65, 66,69> -p4-	11,12, 16,17, 18,20, 41,65, 66,69> -p4-	18,20, 42, 43, 36, 54,65> -p4-	11,12, 16,17, 18,20, 41,65, 66,69> -p4-			67,39, 20> -p5-	**TR**				<67, 39,38, 35,20 -p5-	
													7> -p4-	**PT**				
	13> -p3-		13> -p3-	13,84> -p3-	13,84> -p3-	13,84> -p3-	13,84> -p3-	13,84> -p3-	13,84> -p3-			13> -p3-	13> -p3-	13> -p3-	**SF**		<13 -p3-	<13 -p3-
1> -p1-	1> -p1-	1> -p1-	1> -p1-	1> -p1-	1> -p1-	1> -p1-	1> -p1-	1> -p1-	1> -p1-	1> -p1-	1> -p1-	1> -p1-	1> -p1-	1> -p1-	1> -p1-	**IS**	<1 -p1-	<1 -p1-
	58> -p7-				56> -p6-	56> -p6-	56> -p6-	56> -p6-	56> -p6-			56> -p7-	63> -p4-	79> -p5-			**SN**	
					59> -p6-	59> -p6-	59> -p6-		59> -p6-									**RV**

Figure 8-3: ETCS modes transition table (from SUBSET-026-4)

This is a matrix with the modes listed in the squares that form the diagonal from the top left to bottom right. Each of the other white squares in the same column as a mode represents a potential mode transition to another mode (in the same row as the white square). Each white square lists a reference number and priority level that defines the conditions for the transition. A grey square is a non-permitted transition.

By way of an example, figure 8-4 shows how the transition table in figure 8-3 is applied for the transitions between Staff Responsible (SR) and Full Supervision (FS) modes. This shows that conditions [31] and [32] apply to the transition from SR to FS, and condition [37] applies for the transition from FS to SR. The priority -p6- applies in both cases. Priorities become relevant if several conditions can be fulfilled at the same time which could, without prioritisation, enable more than one mode transition (-p1- is the highest level priority).

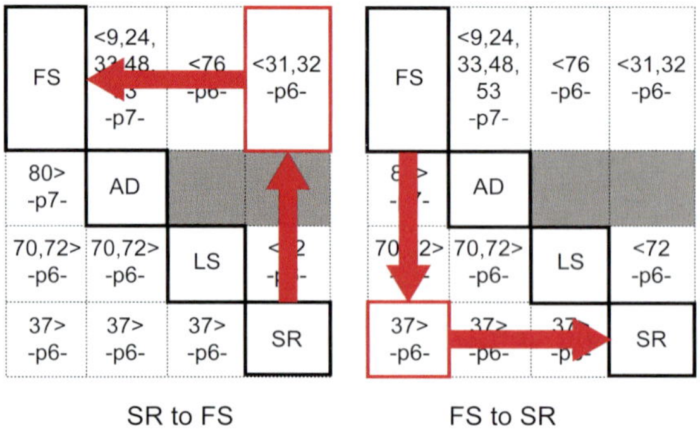

[31]	(movement authority, static speed profile and gradient are all available on board) AND (the train position confidence interval does not overlap any mode profile) AND (ETCS level is 2)
[32]	(movement authority, static speed profile and gradient are all available on board) AND (the train position confidence interval does not overlap any mode profile) AND (ETCS level is 1) AND (no trip order is given by balise)
[37]	(driver selects 'override') AND (train speed is under or equal to the speed limit for triggering the override function) also see {3} detailed description of override procedure

Figure 8-4: Extract from transition table showing transitions between SR and FS modes

References

SUBSET-026-4: System Requirements Specification, Chapter 4 Modes and Transitions

Chapter 9: Operational Procedures

9.1 Operational Procedures

ETCS includes a set of standard procedures necessary for interoperability. These describe the requirements for the interaction of the ETCS on-board with the driver in various situations and are defined in the System Requirements Specification SUBSET-026, Chapter 5. This chapter describes some of these procedures and the associated ETCS design considerations, where the driver and in some cases the control centre operator may be required to interact with the system, namely:

- Start of Mission
- Level transition scenarios (including entry into ETCS and exit from ETCS)
- Change of mode
- Cross border movements
- End of Mission
- Override procedure and Staff Responsible mode

9.2 Start of Mission

The Start of Mission (SoM) procedure is triggered by the driver entering a cab and opening the desk. The exact interpretation of 'opening the desk' varies depending on the type of train/locomotive but usually involves the driver inserting a key and selecting the forward direction of travel.

After opening the desk, the system is designed to prompt the driver to enter data and make selections in a series of steps via the DMI. The SoM procedure itself is complex and contains many options depending on how the system is configured and what the driver selects. Additionally, data entry for national systems and brake tests may also be amalgamated into the SoM procedure.

A typical SoM procedure comprises the following steps:

- Driver identification data entry
- Train running number data entry
- Level selection
- Train data entry
- Mode selection
- Start

The driver identification and train running number steps are self-explanatory, serving to uniquely identify the driver and the train service. The driver simply needs to enter this information when prompted. The other steps are explained in more detail below:

Level selection: The ETCS on-board system remembers the ETCS level from the previous mission and displays this on the DMI during SoM. If this level has been validated by the cold movement detection function (meaning that the train has not moved since the previous mission ended), no action is required. Alternatively, the driver is requested to confirm the previous level or select an alternative level.

Train data entry: A data entry form is displayed on the DMI during SoM for the driver to enter the relevant train data. This data entry form can vary for different train types and different ETCS suppliers. The main types of train data to be entered by the driver are train categories, train lengths and braking percentages (the last of these is only applicable for variable length trains using a particular type of braking model). Any data required for national train protection systems is also entered.

Mode selection: Some operating modes can be entered directly by driver selection on the DMI – these are Non Leading (NL), Shunting (SH) and Supervised Manoeuvre (SM) modes. In the case of NL and SH modes no train data entry is required. For SM mode a preconfigured set of default train data is used and the safe train (consist) length must be available – this length is passed from the train systems to the ETCS on-board, not entered by the driver.

Start: For modes other than NL, SH and SM, the driver, having selected and entered all the necessary information, presses 'Start' for the ETCS to determine the appropriate mode. The next steps depend on the ETCS level. For levels 0, 1 and NTC the on-board determines the mode offered to the driver, while in level 2 it is determined by the RBC. This is explained in the flowchart in figure 9-1.

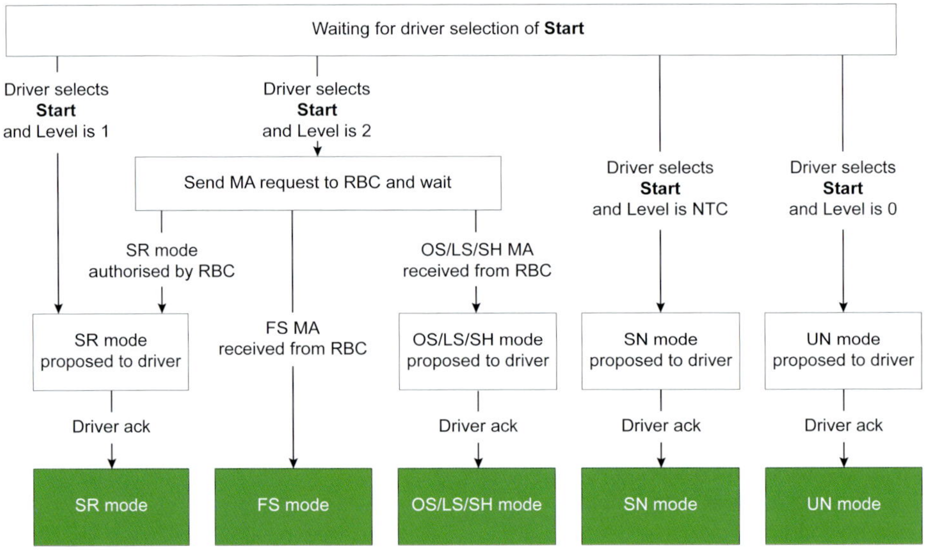

Figure 9-1: Flowchart for start of mission

The mode selection by the RBC in level 2 depends on a number of factors such as whether the on-board position is known, whether a movement authority is available, whether a mode profile is applicable, and whether the train position is overlapping a mode profile area. This will determine which mode can be offered as follows:

SR mode – train position unknown. RBC cannot issue a movement authority.

OS/LS/SH mode – train position known, movement authority with static speed profile (SSP), gradient and mode profile available, and train position is overlapping the mode profile area.

FS mode – train position known, movement authority with SSP, gradient available, and train position does not overlap any mode profile.

Where driver acknowledgement is required following mode selection by the ETCS on-board, this is done by the driver pressing the flashing mode icon on the DMI.

In level 2 there is a useful function known as **Track Ahead Free (TAF)** to enable the train to upgrade to FS mode as soon as possible in situations where the train starts its mission at some distance before the entry point to the next block section, where the movement authority using FS mode would otherwise start. In these cases, the starting mode will be SR or OS (or LS in special cases). If the TAF function is enabled, the RBC can send a 'Track Ahead Free' request to the ETCS on-board, which is displayed on the DMI. If the driver then presses the 'Yes' button next to the 'Track Ahead Free' symbol (to indicate there are no obstructions visible ahead), the ETCS on-board sends a message back to the RBC. This allows the RBC to update the movement authority (provided all other conditions are fulfilled) and send it to the train, which then allows the ETCS on-board to upgrade to FS mode. The TAF function is particularly useful in places where timing-keeping is important for punctuality, for example at terminus platforms where the driver needs to change ends and perform a new Start of Mission.

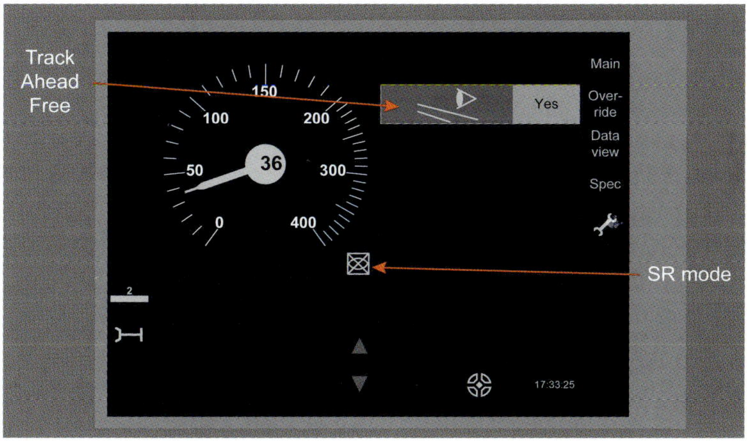

Figure 9-2: Track Ahead Free function on DMI

9.3 Level transition – entry into ETCS area

Refer to chapter 7 for an explanation on level transitions and how they are applied. This section and the following ones build on those general application principles and describes some typical level transition scenarios.

A train travelling towards an ETCS area is likely to be in level NTC (for legacy signalling areas) or level 0 (for unequipped areas). In both cases the train is able to perform a level transition on the move into ETCS level 1 or 2 as described in the following examples. Note that driver acknowledgement is not required for these scenarios because transitions to both ETCS level 1 and level 2 offer the highest level of automatic train protection.

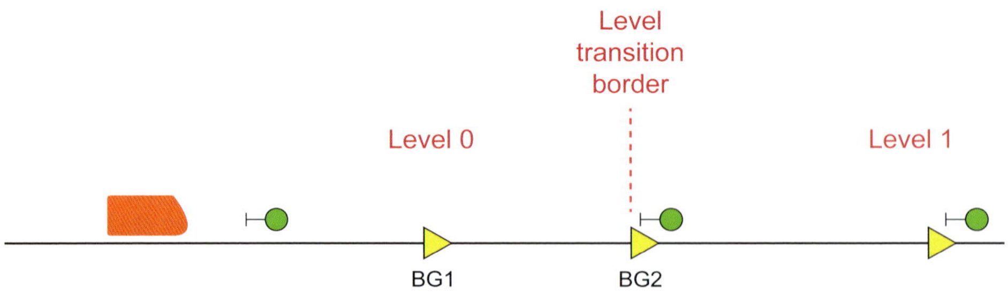

Figure 9-3: Entry into ETCS level 1 area

Figure 9-3 shows an example of a transition from level 0 to level 1. The driver drives according to the lineside signals towards the level transition border. Alongside the track, lineside signs are located informing the driver about the coming level transition to ETCS level 1.

Balise group BG1 provides an announcement of level transition to level 1 and the distance to the level transition border. Provision of this balise group is optional, and where it is provided it can also include a track description into the ETCS area. Once the train has passed BG1, the approaching level transition is shown to the driver on the DMI. Balise group messages valid for level 1 will be accepted by the ETCS on-board system and stored until the level transition border is passed.

Balise group BG2 contains a level transition order together with an MA and track description covering, as a minimum, the distance to the next signal. When BG2 is passed and the ETCS on-board has accepted the information contained in the balise group, the level will be changed to level 1 and displayed on the DMI, and the mode will be changed to FS mode if such information was received from the balise group.

Figure 9-4 shows an example of a transition from level 0 to level 2. In order to make such a transition the ETCS on-board must firstly connect to the RBC. The time to register to the relevant network and set up a communication session with the RBC can be more than a minute. The timings will depend on several factors including the performance of the RBC, the number of trains within the RBC area, the number of transitioning trains, and the radio network itself.

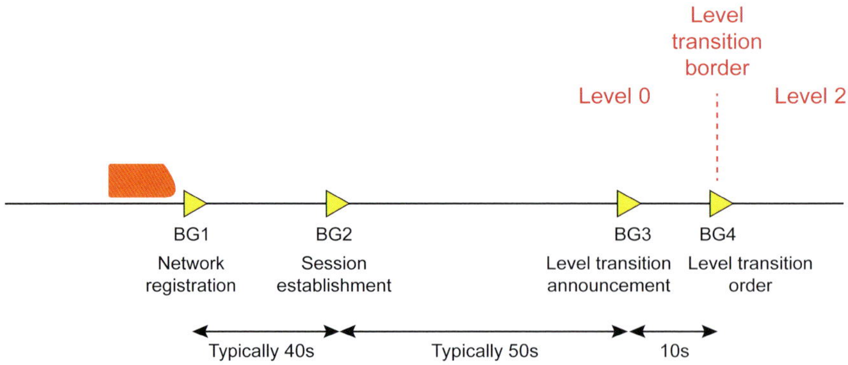

Figure 9-4: Entry into ETCS level 2 area

BG1 contains an order for the train to register to the radio network relevant for the area where the RBC is located (if this information has not already given to the train from other balise groups).

BG2 contains an order to set up a communication session with the RBC controlling the railway network beyond the level transition border at BG4.

BG3 contains the announcement of level transition to level 2 and the distance to the border (the distance between BG3 and BG4). The location of BG3 in relation to the border must allow enough time for the necessary information exchange between RBC and ETCS on-board system, taking into account the highest permitted speed. Note that it is also possible for the RBC to send the level transition announcement over the radio.

BG4 contains the (immediate) level transition order to level 2.

More balise groups may be placed on the approach to the level transition so that the ETCS on-board can send an accurate position report and so receive a movement authority earlier, to avoid any risk of a trip.

A typical scenario for this transition is as follows:

1. The train passes over BG1 and BG2. The ETCS on-board registers its radio(s) with the network and sets up a communication session with the RBC, including the check of compatible ETCS system versions. Once the communication is set up, the ETCS on-board sends train data values and a MA request. The RBC replies by sending acknowledgement of train data along with national values, position report parameters and MA request parameters.
2. The train passes BG3 and the announcement of the transition to level 2 is shown to the driver. The ETCS on-board will send a position report to the RBC, defining train position in relation to the location of BG3. Radio messages will now be accepted by the ETCS on-board system and stored on-board until the level transition border is passed. The RBC will send a MA and track description using BG3 as the last referenced balise group (LRBG) as soon as there is a route set for the train across the border.
3. The train passes the level transition border. Level 2 is indicated to the driver and mode is changed to FS if the MA received from the RBC so allows. If no MA and track description is received before passing the border, the mode will be changed to TR and the emergency brake will be applied by the ETCS on-board.

Note that in some cases, where there is insufficient distance for trains to be able to transition into level 2 (for example, when leaving a yard onto the mainline), it is possible to design a 'level 1 launch'. In these cases, trains transition from level 0/NTC to level 1, and subsequently to level 2 after the session has been established and after the ETCS on-board receives a MA from the RBC.

9.4 Level transition – exit from ETCS area

Trains exiting an ETCS area (in level 1 or 2) will typically transition to level 0 (UN mode) or level NTC (SN mode). In both cases the train can perform a transition on the move provided that the system has information about the signalling status of the area beyond the border. The driver is required to acknowledge the transition either a configured distance on the approach to the border, or at the border itself. This is to make sure the driver is alert to the fact that they are entering an area that has, potentially, a reduced level of automatic train protection. Failure to acknowledge the transition within a set time (5 seconds) will result in a service brake application.

Figure 9-5 shows a typical scenario that is equally applicable to a train in level 1 or level 2.

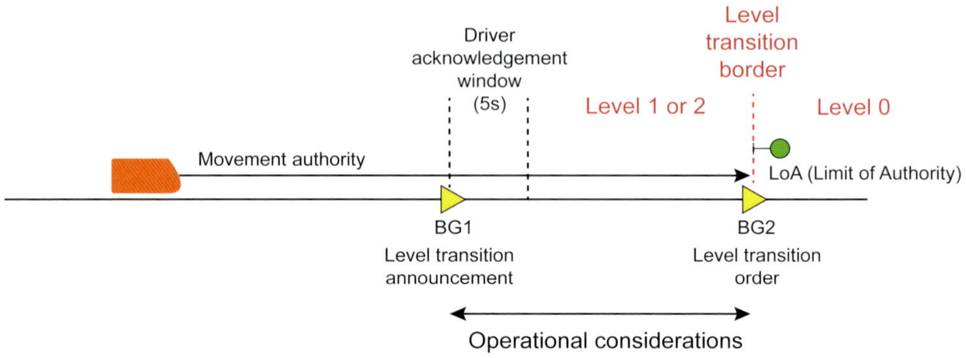

Figure 9-5: Exit from ETCS level 1 or 2 area (using Limit of Authority)

1. The train receives a movement authority that provides a permitted speed at the border location (a Limit of Authority – LoA). To do this the ETCS trackside system needs to acquire the signalling status of the area after the border, for example by determining the status of the entrance signal (shown green in figure 9-5). If the entrance signal is showing a proceed aspect, the LoA permitted speed will be set to a value appropriate to the line ahead of the signal, whereas if the entrance signal is showing a stop aspect, the LoA permitted speed will be set to zero. It is also possible to extend the movement authority into the area after the border instead of providing an LoA at the border. Both methods are permitted and the choice of which one to use will depend on detailed design considerations.
2. The train continues towards the border and at balise group BG1 it receives a level transition announcement along with an acknowledgement window (set at a pre-configured distance before the border or at the border itself). In level 2 this information can instead be sent from the RBC.
3. The driver acknowledges the level transition on the DMI within the pre-configured time (5 seconds).
4. The train passes over the level transition order balise group BG2 and transitions into level 0 or level NTC as appropriate. If the train was in ETCS level 2, then the RBC will send an order to disconnect the radio session after receiving a position report confirming that the whole train has passed over the transition border.

9.5 Mode change initiated by mode profile

The ETCS trackside is able to send a mode profile to accompany the movement authority in levels 1 and 2. A mode profile is expressed in terms of mode type, starting distance, length and speed. The three possible mode profiles are OS, LS and SH. Mode profiles are used for a variety of scenarios such as:

- **OS mode profile:** permitting a second train to enter an occupied block.
- **LS mode profile:** permitting trains to run in an area where the ETCS trackside knows the status of some signals but not all of them.
- **SH mode profile:** permitting a train to enter a shunting area and then move within that area under the driver's responsibility.

In all cases the driver is required to acknowledge the change of mode on the DMI. Failure to do this within the pre-configured time of 5 seconds will result in a service brake application. In all cases the train is able to change mode on the move provided that the speed of the train is less than the speed permitted for that mode.

If the mode profile ends before the end of the movement authority, a train in OS or LS mode will transition to FS without any driver action. If the train is in SH mode, however, it has to be at standstill and requires driver action in order to change mode, either to SB mode or PS mode.

9.6 Mode change to Trip (TR)

The ETCS on-board system will transition to TR mode for a variety of different reasons where a safety intervention is needed to stop the train. Some examples are given below (this is not a full list):

- Exceedance of End of Authority/Limit of Authority
- Balise linking reaction set to Trip
- Loss of safe radio connection (NVCONTACT) reaction set to Trip
- Unconditional emergency stop message is accepted
- Balise order (e.g. trip order, 'Stop if in Shunt', 'Stop if in SR')
- Balise not in list (e.g. not in list of SR balises, not in list of SH balises)
- Balise version number incompatible with ETCS on-board
- Balise group in linking information passed in the unexpected direction

9.7 Mode change initiated by the driver

A change to any of the following modes can be initiated by driver action, either directly by the driver making a selection on the DMI or indirectly as a consequence of the driver's action, for example when the driver closes the desk.

Isolation (IS): This mode is entered by use of a physical switch in the driving cab, which is moved to the isolate position by the driver or a technician. Depending on the design of the ETCS on-board and the installation, the DMI can still be used by other on-board equipment, for example to display speed from an alternative source (an indication that there is no connection with the ETCS on-board may be shown).

No Power (NP): This mode is entered either by the driver moving a physical power switch to the off position, or automatically when the power to the whole locomotive is switched off. Nothing is displayed on the DMI unless it is separately powered and used by other on-board equipment (an indication that there is no connection with ETCS on-board may be shown).

Stand By (SB): This mode is entered automatically when the driver closes the desk. The symbol for SB mode is indicated on the DMI. When the driver opens the desk, SB mode is the default mode until the Start of Mission procedure is complete.

Sleeping (SL): The ETCS on-board equipment in a cab will enter SL mode if the driver opens the desk in a cab elsewhere in the train. This is automatically detected via the train to ETCS interface. The DMI does not show any information in SL mode.

Shunting (SH): SH mode can be selected by the driver via the DMI or ordered by the ETCS trackside.

Passive Shunting (PS): PS mode is entered when the driver selects 'Continue shunting on desk closure' whilst at standstill and then closes the desk.

Post Trip (PT): PT mode is entered following a Trip. To enter PT mode the driver must wait until the train has come to a standstill and then acknowledge the Trip on the DMI.

Non-Leading (NL): This is selected by the driver during Start of Mission or when at standstill in mode SH, SM, FS, AD, LS, SR or OS. The symbol for NL is shown on the DMI. Current train speed and accepted track condition information is shown on the DMI.

Staff Responsible (SR): SR Mode is proposed when the driver requests Override (see section 9.11). In level 1, SR is proposed during Start of Mission after the driver selects Start. In level 2, SR authorisation is sent by the RBC during Start of Mission after the driver selects Start, provided that circumstances permit it. All transitions from SB to SR mode require driver acknowledgement.

Reversing (RV): This mode is only permitted inside designated reversing areas. To change to RV mode the driver moves the train direction controller to reverse when the train is at standstill. Providing that the train is inside a designated reversing area, the ETCS on-board will request the driver to acknowledge reversing via the DMI.

Supervised Manoeuvre (SM): This mode is requested by the driver on the DMI. In level 2, the RBC will authorise SM mode provided that all conditions are met.

Automatic Driving (AD): The driver selects 'ATO engage' on the DMI. This selection can be made at standstill, or when the train is moving providing that braking has not been requested.

9.8 Mode change initiated by ETCS on-board failure

If the ETCS on-board system detects a fault that affects safety, it will transition to SF (System Failure) mode. The determination of exactly which failures affect safety depends on the design of the ETCS on-board system. In SF mode the ETCS on-board will command the emergency brake to be applied, either actively or as a result of stopping a watchdog function or similar. The ETCS on-board system no longer performs any speed supervision, and the emergency brake is permanently commanded to be applied until the ETCS on-board system is disconnected from the brake by the driver or a technician using the physical isolation switch This puts the ETCS on-board into IS (Isolation) mode.

To recover from SF or IS mode requires the disconnection and reconnection of power to the ETCS on-board. This is known as an 'ETCS reset'. If the physical isolation switch has been restored, internal testing takes place and if this is successful the ETCS on-board will enter SB (Stand By) mode. Alternatively, if the failure is permanent the ETCS on-board system remains in SF mode. In this case, the ETCS on-board system must be isolated again (by the driver or a technician) to enable the train to be moved to a repair centre.

9.9 Cross border movements

In ETCS the phrase "cross border" generally means crossing from one country to another, but it can also mean crossing between different regions within a country. Note that in ETCS terminology, a country or region is an area for which a unique value of NID_C has been defined. To enable ETCS cross border movements, consideration must be given to the following factors:

Interlocking-interlocking interface: Interlocking rules and principles vary between railway administrations. ETCS requires that these differences are handled by the trackside system and that the messages sent to the ETCS on-board system are the same on both sides of a border.

Typically, the interlocking-interlocking interface already exists and requires minimal adaptation for ETCS cross border movements.

ETCS level: Different ETCS levels could be used on the two sides of a border, which implies that there is a level transition at the border. Where level 2 is applied on both sides of the border there is no level transition but there will be a RBC-RBC interface as described in chapter 22.

Radio network: To allow for a smooth transition at a border in a level 2 area, the instruction to register to another radio area network as well as to a new RBC may need to be sent to the train well before the border. To achieve this, it is important to have sufficient overlap of radio areas. If this is not possible, an alternative solution would be to install a level 1 area at the border, between two level 2 areas.

System version: The railway administrations on either side of a border may use different ETCS versions. The ETCS system version is sent by the trackside to the ETCS on-board and is checked for compatibility with the on-board system. Checking compatibility of the ETCS version trackside and on-board takes place at each balise as well as when a communication is first established between the ETCS on-board and the RBC. See chapter 14 for further detail on system versions and national values.

National values: A set of national values is defined to be valid for one or several values of NID_C. The ETCS on-board stores these national values so that they are always available, even when the system is powered down (NP mode). If the balise group at a border includes a new set of national values, the previously used national values are replaced with the new values, together with information about which NID_C(s) the new values apply to. If the balise group does not contain a new set of national values, the previously used values are replaced with the default national values and there will be no matching of NID_C until a new set of national values has been received. National values are sent from trackside, via balise or RBC, together with the distance to the start of their validity. When the train reaches the start of validity location, the new set of national values will be used, and as the train passes over each subsequent balise group a check is made of the NID_C to confirm that the existing national values still apply. Note that the default national values are also used when the ETCS on-board system has been modified, for example if a new data configuration or software version has been uploaded.

9.10 End of Mission

End of Mission refers to the situation where the ETCS trackside ceases to be responsible for authorising the movement of a train. For the ETCS on-board, End of Mission occurs when there is a change of mode to Stand By (SB) or Shunting (SH) mode. For example, when a train reaches the end of its journey, the driver closes the desk and the ETCS on-board enters SB mode. All route related information, for example MA and track description, is deleted in the ETCS on-board. If the ETCS on-board is in level 2, the mode change is reported to the RBC in an End of Mission message, including a position report. This will result in a request from the RBC to terminate the communication session, and the ETCS on-board then terminates the session.

While in SB mode with power to the ETCS on, ETCS on-board will perform standstill supervision, and will apply the emergency brake if movement is detected. In SB mode the ETCS on-board will also keep information concerning the position of the train, the current ETCS level (including the table of priority of trackside supported levels) and RBC data. Train data, driver ID and train running number are marked for revalidation when a new Start of Mission is initiated.

If the train is to remain stationary for several hours, for example overnight, the power may be switched off after closing the desk, and the ETCS on-board will enter No Power (NP) mode. The stored train position, current ETCS level (including the table of priority of trackside supported levels) and RBC data are all marked as 'invalid'. On restoration of power, provided that the train is fitted with cold movement detection, the ETCS on-board will use the cold movement status to re-validate this information. If no cold movement has occurred, the information will be updated to 'valid'. If cold movement has occurred, the data will remain marked as 'invalid' and will have to be acquired afresh when a new Start of Mission is initiated. Information that is not affected by cold movement, such as national values, radio network and virtual balise covers, is stored by the ETCS on-board and remains valid even if cold movement has occurred.

9.11 Override procedure and Staff Responsible mode

Inevitably there will be scenarios when a driver has stopped at an End of Authority (EoA) and either no movement authority (MA) can be issued or, in the case of ETCS level 1 or level 2 overlay, the lineside signal cannot display a proceed aspect. Typical reasons for this could be a failed train detection section, a failed switch (points) in the track, or a radio communications failure. In these scenarios, operational procedures accompanied by written orders are required to ensure that a secure protocol is followed between the driver and the control centre operator to allow the driver to override the EoA and proceed in SR (Staff Responsible) mode.

Key elements of this protocol are as follows:

1. Agree the location of the train
2. Establish the reason(s) why the MA cannot be issued
3. Agree the extent of the authority to proceed in SR mode (usually up to a specific signal or ETCS stop marker)
4. Agree the maximum speed to be used and any specific hazards to look out for during the movement
5. Agree when and where the driver should report back to the Signaller

Having completed this protocol (usually recorded by both the driver and signaller completing a paper written order form), the driver presses the 'Override' button on the DMI. This opens the specific override window from which the driver can select 'EoA'. This two-step process helps to avoid accidental selection. On selecting 'Override', the mode changes to SR unless the train was previously the train was in SH mode, or level 0/level NTC. Once in SR mode, the driver may modify the SR mode speed and distance limits on the DMI to those agreed with the control centre operator, provided that the train is at standstill. The SR mode speed and distance limits may also be sent by the RBC.

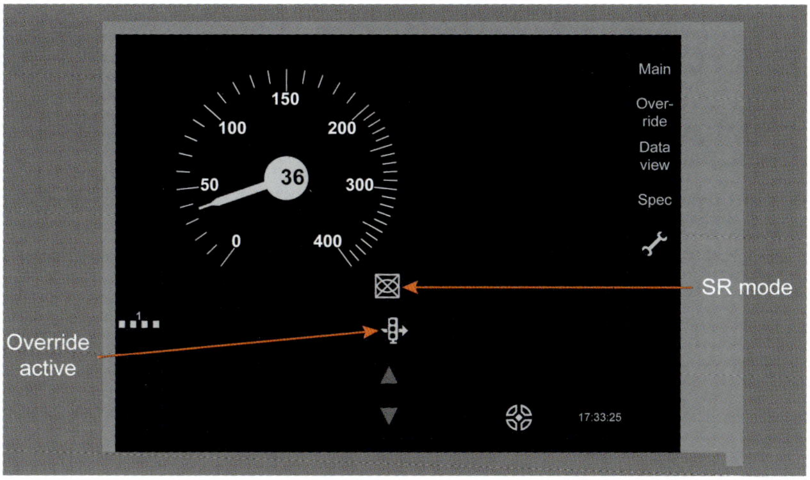

Figure 9-6: DMI showing SR mode with Override EoA active

The maximum extent of the override is limited by time and distance national values. There are many conditions that allow the ETCS on-board to exit the override. For example, if the ETCS on-board receives a valid movement authority, e.g. a level 2 OS (On Sight) movement authority, this will cause it to change mode and exit override.

To reduce the risk of a train in override proceeding beyond the agreed extent of authority, it is desirable to program the balise groups at signals and stop markers with a 'Stop if in SR Mode' data packet. This means that if the driver continues beyond the agreed extent of the authority, the ETCS on-board will enter TR (Trip) mode and apply the emergency brake.

In an ETCS level 2 'no signals' application, the braking distance of a train that has been tripped by the 'Stop if in SR mode' packet becomes a significant factor in determining the location of ETCS stop markers. If a higher SR mode speed is selected this necessitates longer sighting distances and longer braking distances, which will in turn require the stop markers to be set back further from (before) their associated danger points.

References

SUBSET-026-5: System Requirements Specification, Chapter 5 Procedures

Chapter 10: Movement Authority and Track Description

10.1 Introduction

Permission for a train to move is traditionally shown as a signal aspect on a colour light signal. The signal aspect is controlled from an interlocking based on the route that has been requested from the traffic control system, and it is also dependent upon track occupancy by other trains. Signal aspects vary and have different meanings in different countries, and train drivers interpret the meaning of each aspect in combination with their route knowledge and other information such as lineside signs to determine how far and how fast to drive the train.

In ETCS a movement authority is a message defining the distance a train is allowed to travel. Associated with it is a track description. The latter provides information about the route ahead so that the driver and the ETCS on-board can control the train safely. The track description always includes the static speed profile (SSP) and the gradient profile. It may also include supplementary information that may be relevant for some trains and routes such as:

- Axle load speed profile
- Speed restriction to ensure that a given permitted braking distance is sufficient
- Track conditions, e.g. a section without electric power for traction, non-stopping area, radio hole
- Areas where reversing is allowed
- Adhesion factor
- Route suitability data, e.g. loading gauge, traction system, axle load

The movement authority and track description are processed by the ETCS on-board (also making use of relevant train data) to calculate the speed profile applicable to the train, the location where braking should commence in order to stop at the end of the movement authority, and to generate the driving information on the DMI.

In level 1 movement authorities are sent from the track-mounted balises, based on the signal aspect and other information such as point (switch) direction if there is more than one route from a signal. In level 2 and where a 'centralised LEU' architecture is used in level 1, the movement authority is derived directly from the interlocking.

In level 1 the distance and direction information in the movement authority and track description is always given with reference to the balise group giving the information, while in level 2 the radio messages include a reference to the unique identity of a specific balise group in the variable NID_LRBG (last relevant balise group). Further information on location referencing in ETCS is provided in chapter 17.

10.2 End of Authority and Limit of Authority

In most situations the movement authority and track description contain all the information required for the train to reach the 'End of Authority' (EoA) location, where the train is required to stop. This is called 'distance-to-go' signalling – see figure 10-1.

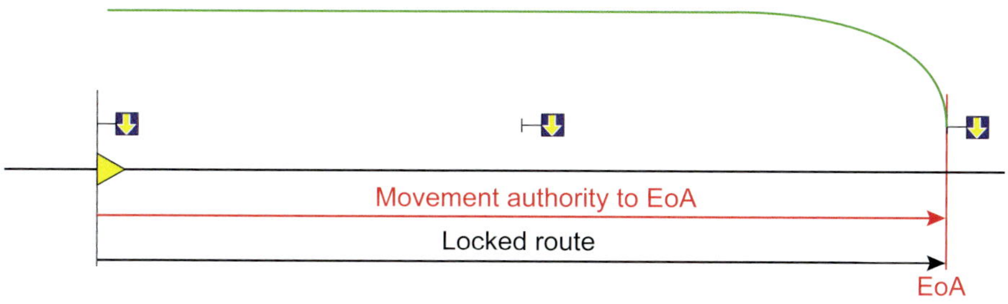

Figure 10-1: Movement authority to End of Authority (EoA)

However, there are some situations where this is not possible, and in these cases the train can be sent a 'Limit of Authority' (LoA) location (which is part-way along the route, before the end of authority location) and a speed that must not be exceeded at that location. This is called 'speed step' signalling. This might apply in a level 1 application where it is known that the signal ahead is displaying a proceed aspect, but the precise route beyond that signal is unknown. The expectation is that the train will receive an updated movement authority and track description before it reaches the limit of authority location. If no update is received when the train passes the limit of authority, the ETCS on-board will enter TR (Trip) mode and apply the emergency brakes. A limit of authority can also be used on the approach to a level transition, for the example in figure 10-2 where a national system will take over control from ETCS beyond the limit of authority location.

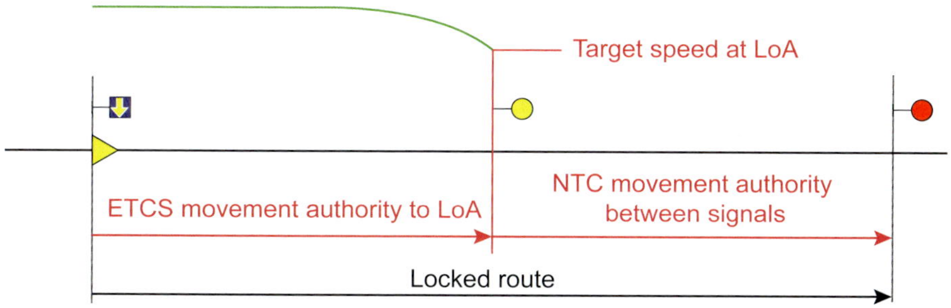

Figure 10-2: Movement authority to Limit of Authority (LoA)

10.3 Supervised location and release speed

An ETCS movement authority can also define a 'Supervised Location' (SvL) beyond an EoA. Where this is provided the emergency brake intervention is calculated to stop the train before the SvL instead of before the EoA. This allows the calculation of a 'release speed', which is slow enough to ensure the emergency brake can still stop the train before the supervised location if the train passes the end of authority. The purpose of the release speed is twofold:

- In level 1 release speeds are essential to allow a train that has stopped at a signal to move forward over the balise group at the signal to receive an updated movement authority when the signal changes to a proceed aspect.
- In level 2 a release speed avoids the situation where inaccuracy in the ETCS distance measurement (see chapter 17) could mean that the train has to stop before reaching the

physical location of the EoA. This can cause operational difficulties; for instance, in the case of a long train, the points (switches) at the entry to a station might be blocked by the last vehicles of the train.

The distance from EoA to SvL can be set to zero, or defined as the distance to the 'danger point' or the 'end of overlap' where these concepts are used in the interlocking system (see figure 10-3). The supervised location is defined as:

- the end of overlap (if one is provided, and before it times out – see section 10.5 below)
- if not, the danger point (if any)
- if not, the end of authority (in which case no release speed is possible)

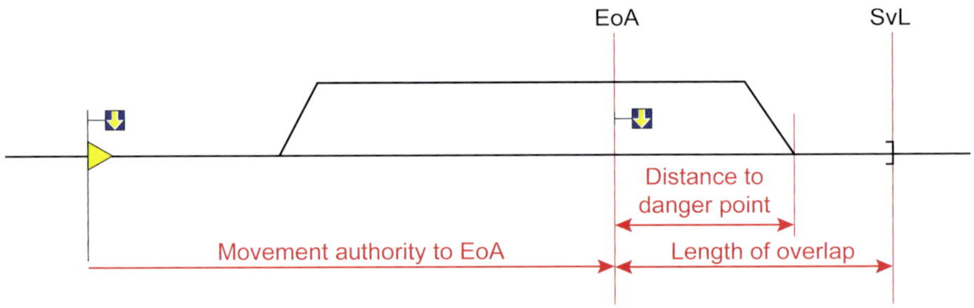

Figure 10-3: Danger point and end of overlap

The release speed can be calculated either by the trackside ETCS or on the train. If calculated trackside, it will take into account the worst braking performance of all the trains running on the line and the braking distance to stop when running at the release speed. This braking distance from the EoA must not exceed the distance to the SvL.

If the release speed is calculated on-board, the ETCS trackside equipment must give the distance from EoA to SvL as part of the movement authority. The ETCS on-board can then use the braking performance of the train to calculate the release speed that is appropriate for stopping before the SvL.

10.4 Withdrawal of movement authority

When choosing the length of a movement authority to be given to the train, there is a conflict between flexibility in the use of the track and the smooth, uninterrupted movement of a train. By using longer lengths, for example two or more signal sections, train operation will be more efficient as it avoids unnecessary braking. But the risk of something happening after the movement authority has been issued that would require a change to the locked route (and therefore a change to the movement authority) increases with longer movement authority distances. In normal operation these route changes are inevitable. ETCS level 2 includes features that allow route changes to be managed more safely and efficiently whereas ETCS level 1 replicates the route release timers that are used in traditional signalling.

To protect a train for which a movement authority has already been issued, and a route cancellation request is subsequently made by the traffic control system, a timer in the interlocking may enforce a delay before the route locking is released. When the interlocking receives the cancellation request, the movement authority is withdrawn (signal replaced to danger) and the timer is started. When the timer has timed out, the route locking is released and the track can be used for another train movement if the conditions for that new movement are fulfilled. The timer duration is sufficient to ensure that a train with the worst braking performance comes to a stop before the signal whose aspect has been reset to danger (red), taking into account the location at which the driver can first see the change of aspects of preceding signals and start braking the train. If the train passes the signal that has been reset to danger before the timer times out, the interlocking will maintain the locking on the original route and prevent a new route from being set and locked.

To synchronise between the trackside and the ETCS on-board in these situations, the movement authority can be split into several sections, and each section can be given a section time-out value and section time-out stop location. When a movement authority is received, the ETCS on-board will start a timer for each section for which a time-out is specified (or reset the timer to zero if this is an update to a previously received movement authority and the timer is already running). If any section timer reaches its specified time-out value before the train reaches the corresponding section time-out stop location, the ETCS on-board will shorten the movement authority to the start of that section. This ensures that the ETCS on-board does not retain a movement authority for a route that has been cancelled, and which has timed out in the interlocking. The section time-out values are chosen to fit with the time-out values used in the interlocking system to synchronise trackside and on-board supervisions.

In level 2 a request for the shortening of the movement authority can be sent from the RBC to the ETCS on-board when a route release is requested. Once a positive reply is received from the train to confirm that it can stop safely at the proposed new end of authority using a service brake application, the locked route can be released in the interlocking and used for another train without waiting for the usual time-out delay. This is known as 'cooperative movement authority shortening'.

Level 2 also provides two types of emergency stop message that can be sent from the RBC to the ETCS on-board. An 'unconditional emergency stop' causes the train to immediately enter TR (Trip) mode and the emergency brake is applied. A 'conditional emergency stop' specifies a location and will be rejected by the ETCS on-board if the train has already passed that location. If the train has not yet reached the specified location, the emergency stop is accepted, and the existing movement authority is shortened to end at this location. Braking of the train to stop at the new end of authority is then managed by the driver and the ETCS on-board in the usual way.

The emergency stop facility will not be available if there is a loss of radio contact with the train. To mitigate this risk, the radio link is supervised, and if loss of radio contact is detected, a service brake application or trip can be applied (the time to trigger this, and the action taken, are defined as national values).

10.5 Time-out of overlap

In interlocking systems that incorporate overlaps, timers can also be used for the last train detection section at the end of a route, to manage the release of an overlap. Once a train has entered this last track based train detection section, the timer is started and when it has timed out, it is assumed that the train has come to a standstill at the signal and the risk that the train will pass the signal at danger is negligible. The overlap can then be freed and a route for another movement in the overlap area can be set up and locked.

To synchronise between the trackside and the ETCS on-board in this situation, the last section within the movement authority can be defined as the 'end section'. Where an overlap exists, the movement authority will define the following information for the end section:

- distance from end of section to end of overlap
- time-out value
- distance from overlap timer start location to end of section
- release speed related to overlap

When the train enters the end section, the ETCS on-board starts the overlap timer. When this timer reaches the timeout value or the train comes to a stand, the overlap information is deleted and is no longer taken into account in deriving the SvL as explained in section 10.3. This ensures that the ETCS on-board supervision of the train movement allows for the overlap timing out in the interlocking which could allow another train movement in the overlap area.

10.6 Update of movement authority

As a train progresses along a route, the movement authority available to the train will be progressively extended as the traffic control system sets further routes along the planned path and track sections ahead become vacant as a result of the movement of preceding trains. This is reflected on the DMI based on an updated movement authority sent to the train from balises (level 1) or from the RBC (level 2).

In level 1 the updated movement authority will be received by the train when it passes a balise linked to a signal that has changed aspect (either at the signal or at a preceding infill location).

In level 2 the updated movement authority requires an exchange of data messages between the RBC and the ETCS on-board, and there are two options for how this takes place. A movement authority request can be generated by the on-board system when it identifies that the train is approaching the location where the braking curve begins for the end of the current movement authority. Alternatively, the RBC may send a movement authority to the train without it being requested by the on-board system, for instance when a new route is set or the preceding train has moved, and the current movement authority can therefore be extended. Where level 2 is applied in a moving block configuration, the movement authority can be updated more frequently, based on train position reports from the train in front.

In most cases the updated movement authority will be from a new location reference – either the balise group sending the message (level 1) or the 'last referenced balise group' defined in the movement authority (level 2). The currently supervised movement authority and track description will be replaced by the new information from the new location reference onwards. In level 2 the RBC takes into account possible inaccuracies in train position information and ensures that the new movement authority information covers the whole of the route set, including making allowance for uncertainty in the exact position of the train.

References

SUBSET-026-3: System Requirements Specification, Chapter 3 Principles

Chapter 11: Speed Supervision

11.1 Types of speed monitoring

The ETCS on-board continuously calculates the safe speed for the train based on the movement authority, track description, operating mode and train characteristics. It displays this information on the DMI and then monitors the actual train speed to determine warnings to the driver and interventions to apply train brakes if the safe speed is exceeded.

There are three main types of speed and distance monitoring, namely:

- **Ceiling speed monitoring** – against the maximum permissible speed that is applicable at the current location on the track.
- **Target speed monitoring** – against a braking curve, to ensure the train arrives at a target location at the correct speed or stops at an end of authority.
- **Release speed monitoring** – against the release speed that is applicable as the train approaches an end of authority.

Each has its own set of supervision triggers, reactions, conditions and DMI indications (symbols, colours, sounds).

11.2 Speed Envelope

There are many reasons to limit the maximum permissible speed for trains, e.g. track curvature, switches and crossings, bridges, tunnels, and the weight bearing capability of bridges and other structures that support the railway. In addition, the vehicle itself will have a maximum permissible speed. The maximum permissible speed may be limited in specific operational scenarios, such as driving on-sight or when performing shunting movements. There may also be a need for temporary speed restrictions that apply due to adverse weather conditions, track defects or locations where staff are working on an adjacent track.

There are eleven static speed restriction categories defined in the ETCS specifications. Some of these apply to specific locations on the track, others apply continuously, and some apply to specific ETCS on-board modes. A location-specific speed restriction can apply for just a short length of track or for longer distances.

1) **Static Speed Profile (SSP)**. The SSP is defined in the ETCS trackside. It is mandatory information and the ETCS on-board will not accept a movement authority without a corresponding SSP and gradient profile. The SSP is split into elements, each with a start location and speed value. The ETCS trackside always sends a basic SSP category and may send additional SSP categories for each element if required. In the case of multiple SSP categories, the ETCS on-board will evaluate and select the appropriate SSP category based on the type of train (as defined during train data entry). If the ETCS on-board does not receive a SSP category for the defined train type, the basic SSP category will be used.
   Specific SSP categories are split into two types:
   - Cant deficiency SSP categories (e.g. for tilting trains that are able to run at higher speeds on curves)
   - Other specific SSP categories based on international train categories (e.g. for freight trains that must run at lower speeds than passenger trains)
2) **Axle load Speed Profile (ASP)**. The ASP is defined in the ETCS trackside. Limitations for different axle loads can be sent to the train, and the ETCS on-board then selects the

applicable speed profile based on the current axle load (entered as part of the train data). Axle-load speed profile is optional and will be sent from trackside only where required.

3) **Temporary Speed Restriction (TSR)**. TSRs are defined in the ETCS trackside when required and apply to all trains. In level 2 several packets with one TSR each can be sent in a single radio message to a train, although normally only one TSR is sent at a time. Each TSR can be given an identity to be used if the TSR needs to be revoked before the train reaches the affected area. A TSR can also be defined as 'revocable', which means that a TSR transmitted by a balise can be revoked by the RBC for a train operating in level 2. In level 1 TSRs are usually applied using portable balise groups that have to be programmed with the relevant data and installed on the track on the approach to the speed restriction. This is a significant disadvantage compared to level 2 where TSRs can be applied from a terminal in the control centre and sent over the radio from the RBC.
4) **Signalling-related speed restriction**. This applies only in level 1 where the trackside signal aspects convey speed information to the driver. The speed value corresponding to the signal aspect is sent as part of a level 1 movement authority and applies from the balise group at the signal location.
5) **Mode-related speed restriction**. These are defined for each ETCS mode, in the form of national values applicable to each railway administration. If no such national values are received from the ETCS trackside, default values specified in the standard are used. There are three exceptions to this:
   a. For OS, LS, and SH modes the speed limit can be given by the ETCS trackside. This will prevail over the national value and default value.
   b. For RV and SM modes there is no national value or default value for the speed limit. The speed limit is always given by the ETCS trackside.
   c. For SR mode the speed limit can be entered by the driver. This will prevail over the national value and default value.
6) **Train-related speed restriction**. This is defined on-board as part of the train data, and can either be fixed for a certain vehicle or vehicle type or entered by the driver during the Start of Mission procedure. For a variable formation train, the maximum train speed is set to the lowest value allowed for any individual vehicle in the train consist.
7) **Level crossing speed restriction**. A movement authority from the ETCS trackside can contain packets of information for level crossings. If the level crossing is 'unprotected' (i.e. warning devices for road traffic are not provided or have failed), this will include a speed restriction that applies over the level crossing.
8) **Override function-related speed restriction**. This speed restriction is applied when the override function is active. This is specified as a national value or a default value in the same manner as the mode-related speed restrictions. Typically, this will be set to the same value as the SR mode speed restriction.
9) **Speed restriction to ensure a given permitted braking distance**. In some situations (e.g. short signal sections), the infrastructure manager may specify a maximum permitted braking distance. In this case the ETCS trackside sends the permitted braking distance to the ETCS on-board, which then calculates a speed restriction to ensure the train can stop within this distance based on its own braking performance.
10) **STM maximum speed**. This allows an on-board specific transmission module (STM) to define the permissible speed to be supervised by the ETCS on-board at the level transition border from ETCS to a national train control system (NTC), in order to have a smooth speed transition from ETCS to the NTC. The value may be pre-defined in the STM (e.g. the maximum permissible speed allowed by this NTC), or based on information received by the STM from the NTC trackside equipment on the approach to the level transition border.

11) **STM system speed**. The 'STM system speed', together with the 'STM system distance', allows an on-board STM to request a given train speed at a given position before the level transition border, in order for the STM to be able to detect its national trackside.

The static speed restriction categories are completely independent of one another, meaning that a change in one speed restriction category at a location or an ETCS mode change cannot affect any other speed restriction category.

The ETCS on-board will continuously evaluate the 'most restrictive speed profile' (MRSP) applicable to the train on each part of the track, taking the lowest permissible speed of all the applicable static speed restriction categories. A speed decrease is supervised when the calculated location of the front of the train reaches the reduction in the speed profile. When the front of the train reaches an increase in the speed profile, the speed can be increased again. For speed restriction types SSP, ASP and TSR, the ETCS trackside can specify that the speed increase is delayed until the rear end of the train has left the restriction. Figure 11-1 shows an example of how the MRSP is calculated where the SSP is specified to apply until the rear of the train has left the restriction and a TSR only applies to the front of the train.

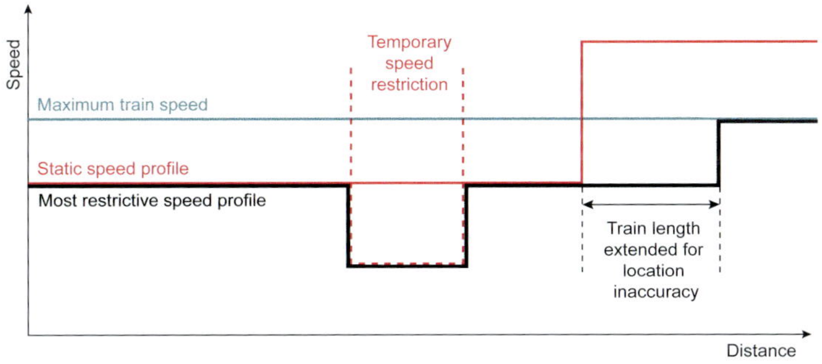

Figure 11-1: Calculation of most restrictive speed profile

Speed supervision is always based on pessimistic assumptions of the train location accuracy. A reduction in speed is based on the **maximum safe front end** and an increase in speed on the **minimum safe front end** or **minimum safe rear end** location, depending on whether the increase is allowed as soon as the front of the train has passed the location or must be delayed until the rear of the train has passed. See chapter 17 for details of how these are derived by the ETCS on-board.

11.3 Ceiling speed monitoring

Using the most restrictive speed profile (MRSP), the ETCS on-board monitors the current train speed against the allowed speed for the current location (the permitted ceiling speed). If the ceiling speed is exceeded by more than:

- the warning margin (4 km/h to 5 km/h depending on speed), an audible warning is issued to the driver.
- the service brake intervention margin (5.5 km/h to 10 km/h depending on speed), a service brake command is issued.

– the emergency brake intervention margin (7.5 km/h to 15 km/h depending on speed), an emergency brake command is issued.

The audible warning is withdrawn when the speed has been reduced below the ceiling speed.

Service braking is revoked automatically by the ETCS on-board when the speed has been reduced below the ceiling speed.

A variable Q_NVEMRRLS (a national value) controls whether emergency braking may be revoked only when the train has come to a stand, or after the speed has been reduced below the ceiling speed. In both cases emergency braking is revoked automatically by the on-board system.

11.4 Target speed monitoring

Each speed reduction in the MRSP is defined by a 'target', comprising a target location (which is the commencement of the restriction) and a target speed to which the train must decelerate before reaching the target location. The ETCS on-board will continuously supervise the actual train speed against these targets, and also against other relevant targets that are additional to the MSRP such as the Limit of Authority (LoA), the End of Authority (EoA), and the Supervised Location (SvL). The ETCS on-board will re-evaluate its list of targets when any of the underlying elements change – for example, if a new movement authority is accepted or if the maximum safe front end of the train has passed a target location.

The current (estimated) speed of the train is monitored with respect to dynamic intervention curves (supervision limits) that are derived from the targets being approached. The ETCS on-board continuously checks that it is the most restrictive target against which the train speed is being supervised. A change in current train speed could result in a change of the most restrictive target.

Figure 11-2 shows an example where Target 2 takes priority over Target 1 as the most restrictive target.

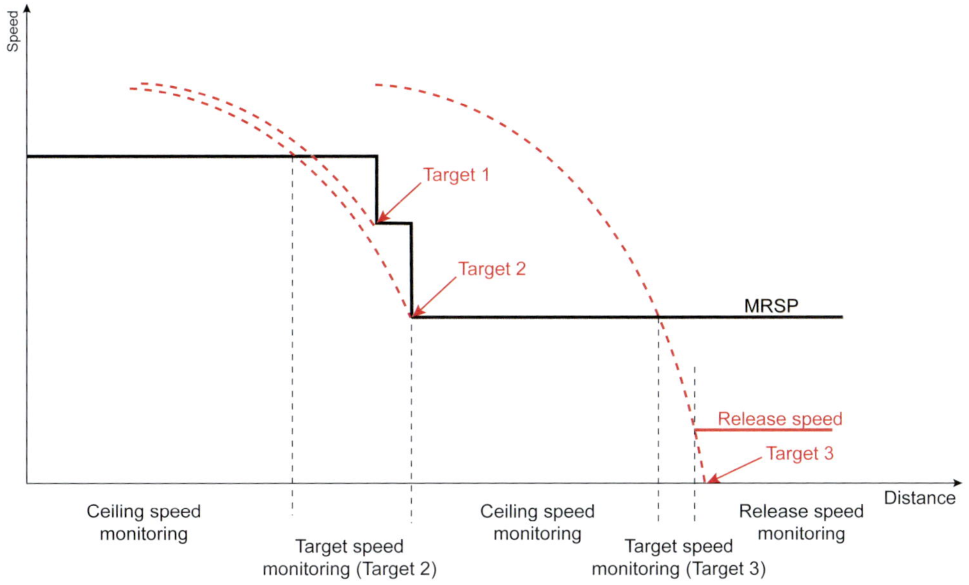

Figure 11-2: Example of speed monitoring to most restrictive target

In the target speed monitoring, a set of curves are defined:

The **Indication (I) curve** defines the speed/location where the ETCS on-board will start to display the required speed reduction and distance to the target location on the DMI.

The **Permitted speed (P) curve** represents the highest speed at which the train can be driven without causing a warning or brake intervention. It can be therefore regarded as the optimal speed to drive the train to achieve a minimum journey time.

The **Warning (W) curve** determines the train speed at which the ETCS on-board will give a warning indication to the driver on the DMI.

The **Service Brake Intervention (SBI) curve** determines the train speed at which the service brake will be requested in the event that the driver does not respond to the warning indication. The purpose of the SBI curve supervision is to serve as a first line of intervention, preventing the train from exceeding the emergency brake intervention curve (see below).

The **Emergency Brake Intervention (EBI) curve** determines the train speed at which the emergency brake will be requested. The purpose of the EBI curve supervision is to guarantee that the train will respect the speed limit at the target location (within the tolerance defined for ceiling speed or for release speed).

Where a supervised location (SvL) has been defined beyond the EoA, the EBI curve will target the SvL and the other curves will target the EoA.

The calculation of these curves is one of the most complex functions in the ETCS on-board and is described in chapter 12.

11.5 Release speed monitoring

Release speed monitoring may be required for two reasons. Firstly, there may be locations where the driver is required to stop close to the end of authority (EoA), for example, on the approach to buffer stops. Without a release speed, the train may be forced to stop short of the EoA due to positional inaccuracy. Secondly, in level 1, the release speed is necessary for the train to pass over the balise located at the signal when the signal changes to a proceed aspect, enabling the ETCS on-board to receive the new movement authority information.

In release speed monitoring the emergency brake intervention supervision limit is equivalent to the release speed. There are no other speed supervision limits. However, if the **minimum safe front end** of the train does pass the EoA, this will result in a trip, causing the train to be brought to a stop (the actual implementation of this is complex and depends on the level). Similarly, if the speed of the train exceeds the release speed the ETCS on-board will immediately apply the emergency brake and stop the train. This can only be revoked when the train has stopped.

11.6 Indications on the DMI

The driver machine interface (DMI) provides the driver with all the information needed to manage the speed of the train within the speed supervision limits. The current permitted speed is always shown as a 'hook' on the outside of the circular speed gauge. This is coloured dark grey during ceiling speed monitoring.

During target speed and release speed monitoring, the outside of the speed gauge is coloured yellow between the current permitted speed and the target speed; the current speed pointer is also shown in yellow, and a vertical bar on the left shows distance to the target location.

When the permitted speed is exceeded, there is a short audible alert and the pointer and circular speed gauge in excess of the permitted speed is highlighted in orange. When the warning speed is exceeded, there is a continuous audible warning.

In Automatic Driving (AD) mode, the indication and warning pointer and circular speed gauge colours are white (instead of yellow and orange), and the sounds are supressed because the driver is not required to take action.

When a brake intervention occurs, the colour of the pointer and speed in excess of the permitted speed changes to red.

Some examples of DMI indications for the three types of speed monitoring are given in figure 11-3.

<table>
<tr>
<td>
</td>
<td colspan="3">Circular speed gauge grey during ceiling speed monitoring.

Permitted speed shown in dark grey.
Speed hook represents permitted speed supervision limit.</td>
</tr>
<tr>
<td>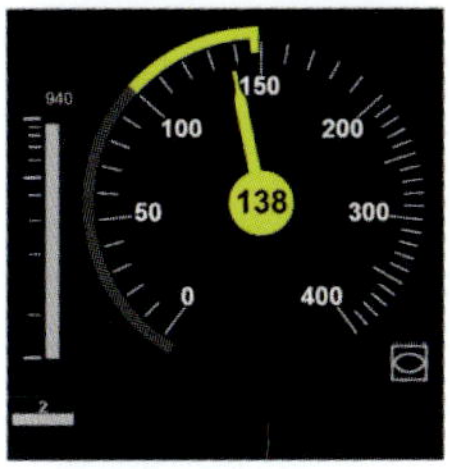

Indication speed in Yellow.
Permitted 148 km/h
Target 100 km/h</td>
<td>

Warning speed in Orange.
Permitted 135 km/h
Target 40 km/h
(Driver has exceeded the permitted speed)</td>
<td>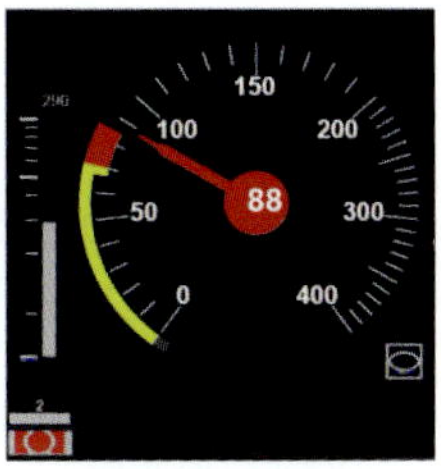

Brake Intervention in Red
Permitted 70 km/h
Target 0 km/h
(Service brake followed by emergency brake)</td>
<td>Circular speed gauge yellow during target speed monitoring.
Speed hook represents permitted speed supervision limit.</td>
</tr>
<tr>
<td>

Target speed monitoring
Permitted 78 km/h
Release 30 km/h
Target 0 km/h</td>
<td>

Release speed monitoring
Release 30 km/h
Permitted 12 km/h
Target 0 km/h</td>
<td colspan="2">Circular speed gauge before and during release speed monitoring.
Release speed is indicated as a numerical value bottom left and a grey strip within the yellow circular speed gauge.
Note that once the train is under the release speed the yellow permitted speed based on stopping at the target location is still shown, but the driver does not need to follow it.</td>
</tr>
</table>

Figure 11-3: Speed monitoring indications on the DMI

References

SUBSET-026-3: ERTMS/ETCS System Requirements Specification, Chapter 3 Principles

Chapter 12: Braking and Performance

This chapter explains the ETCS braking curves and how they play a critical role in determining the safety and performance of the system. It also includes:

- Background information about train braking systems
- Consideration of the problem of low adhesion because this is such a significant factor in the design and implementation of signalling and train control systems
- A description of how ETCS headway is calculated and how this is influenced by other factors in addition to the braking curves

12.1 Train braking systems

A train's brakes are applied either in graduated steps ('brake position'), or as a percentage of the full service brake application, ranging from 'normal service braking' to 'full service braking', and then lastly 'emergency braking'. Despite its name, the deceleration rate for the emergency brake is not necessarily any greater than that of the full service brake, but it is applied by a 'fail-safe' method which is independent of the service brake control.

The overall safety of the railway relies on the ETCS being able to command the emergency brake in due time, although ETCS also includes an option to command full service brake (to avoid nuisance emergency brake applications), and another option to provide guidance to the driver (via indications on the DMI) to encourage earlier and more gradual braking towards a speed reduction or end of authority. These are described in detail in section 12.5.

To monitor a train's speed effectively, and to intervene to slow the train when necessary, the ETCS on-board has to model the train's braking system accurately, taking into account both rolling stock and track characteristics. Key rolling stock factors are the type of brake and the method of actuation which influence the brake deceleration rate used in the ETCS on-board model.

On almost all trains the primary brakes use pressurised air stored on each vehicle to apply a mechanical friction brake on each wheelset (either tread brakes or disc brakes, or both). The friction brake is the backstop for emergency brake applications, sometimes supplemented by other special brakes. For service brake applications, freight trains most commonly use friction brakes, whereas modern passenger trains more commonly use a mix of friction and special brakes (usually regenerative brakes, as described below).

The method of brake actuation can influence the brake reaction times and brake build-up times used in the ETCS on-board model.

Air brake: In an air-actuated system, the friction brakes are applied by the ETCS on-board operating an electromagnetic valve to release the pressure in the air pipe that runs the length of the train. The loss of pressure causes local valves on the vehicles to operate, releasing stored air into the brake cylinders. The brake cylinders then actuate the friction brakes. This system is used when a locomotive is hauling vehicles (passenger or freight) with no electrical connection along the train.

One of the significant challenges with the air brake is that it relies on the drop in air pressure propagating through the air pipe. This can take more than 10 seconds for a long freight train.

Electro-pneumatic brake: This is a variant of the air brake actuation system, which uses electrical wires that run the length of the train instead of an air pipe. When the ETCS on-board commands braking, the power to these wires is removed. This causes brake relays on each vehicle to de-energise, which in turn causes the local valves to operate, releasing stored air into the brake cylinders in the same manner as the air brake system. This system is generally used for fixed formation or multiple unit passenger trains.

Serial data bus: Modern passenger trains use serial data buses that run the length of the train as part of the distributed train management system for both safety and non-safety controls and for status information. Distributed logic modules on the vehicles control the brake and perform functions such as blending different brake types. The ETCS on-board can use this serial data bus interface for all its interfaces with the train's systems, including the emergency and service brakes, and the special brakes if they are provided.

To achieve the highest level of safety integrity for the emergency brake function, it is a mandatory requirement that the ETCS on-board uses hard-wired outputs (or hard-wired in combination with the serial data bus) to command the emergency brake.

12.2 Special brakes

The use of special brakes to supplement friction brakes has various benefits, depending upon their type. They can limit the wear and tear associated with friction brakes, offer energy saving possibilities, or reduce the reliance on rail/wheel adhesion for slowing the train. These special brakes are classified as follows:

Regenerative brakes (electrodynamic): On an electric train the traction motors can act as generators, rotated by the motion of the train and feeding electrical energy back to the traction supply system (which acts as a load). This has the effect of providing a retarding force at the wheelsets which slows the train. Electrodynamic brakes are often only effective when the train is above a certain speed, and the traction system line voltage is low because other trains or station systems are drawing energy. Alternatively, instead of feeding power back to the traction system the train may be fitted with batteries to store the energy or resistors to dissipate it (rheostatic braking).

Eddy current brakes: Electromagnets on the train are energised during braking to induce eddy currents into axle-mounted discs or into the rail itself. These eddy currents create an opposing magnetic field that creates a braking effect without any mechanical contact.

Magnetic shoe brakes: Brake shoes containing electromagnets are lowered onto the rail and energised, creating magnetic attraction force with the rail that causes a friction braking effect which is independent of wheel-rail adhesion.

On a train fitted with special brakes, the brake control system will process the brake demand from the driver or ETCS and control the special brakes and the friction brakes to give the required level of braking, with the optimum 'blend' of different brake types to use available adhesion, maximise energy recovery and reduce friction brake wear.

ETCS interacts with the special brakes in two ways:

- Special brakes can interfere with the correct operation of lineside equipment so they may need to be inhibited in certain areas. This inhibition can be managed via the ETCS on-board using the track information associated with a movement authority.
- ETCS may need to take account of whether the special brakes on the train are operational in calculating the braking curves.

12.3 Use of service braking in ETCS

If the driver exceeds the permitted speed and fails to respond to the warning when approaching a target, ETCS will intervene by making a service brake application to bring the speed under control. This avoids the need for an emergency brake application, which is undesirable except in an emergency stopping situation because of the risk of injury due to high deceleration (especially for standing passengers), the wear and tear on the friction brake components, and the consequential operational delays caused by stopping the train (enforced by the ETCS on-board) and the fact that on many trains there is an extra enforced time delay between the train stopping and the brake release being enabled.

By contrast, service brake applications are smoother, result in far less wear and tear, and they are revocable (meaning that they can be released as soon as the train speed is back under control).

The difficulty with using the service brake in target speed monitoring is that an additional intervention curve has to be inserted between the warning curve and the emergency brake intervention curve – see figure 12-1. This means that the driver is advised via the warning curve to start braking slightly earlier than is really necessary, to avoid an ETCS commanded service brake intervention.

12.4 Impact of low adhesion

Railways generally rely on friction between the train's steel wheels and the railhead to accelerate and decelerate, although there are some exceptions such as the rubber tyred vehicles typically used for automated people movers, and monorail systems. When the level of friction is reduced due to the presence of contamination and moisture on the railhead, the train experiences low adhesion, causing wheels to slip or slide. This can be a serious problem for conventional trains, whether using ETCS or not.

Railways use various methods to reduce the likelihood and consequences of low adhesion. Train-based mitigation measures include:

Wheel slide protection (WSP): Wheel slide is caused when the brake force exerted by the train wheel exceeds the reaction (friction) force exerted by the rail. WSP systems use sensors to monitor the speed of each axle. They can detect individual axle slide, as well as 'all-axles' slide. On detection of wheel slide, the WSP system reduces and reapplies the brake effort on a per-axle basis so that rail-wheel grip can be re-established.

Train-based sander: Automatic sanders can be triggered by the WSP system (where provided) to deliver sand to the wheel-rail interface on a number of wheelsets on the train. The sand helps the wheels to restore grip in low adhesion areas. On some trains an emergency sander may also be fitted, activated by the driver.

Special brakes: As mentioned previously, trains may be fitted with devices such as magnetic shoe brakes that do not rely on wheel-rail friction.

Infrastructure managers also have a major role to play with measures such as:

Vegetation management: In the autumn, falling leaves from trees are a major cause of low adhesion, and controlling their growth on the lineside gives major improvements.

Track-based adhesion improvers: Traction gel applicators can be mounted on the track in known low adhesion areas. When trains pass over the applicator, the adhesion-improving gel is spread onto the rail over the affected area.

Rail head treatment: Special trains can be used to remove railhead contamination by spraying a water jet or applying adhesion-improving gel onto the rails.

Low adhesion zone signs: These warn train drivers to brake more cautiously in areas of known low adhesion.

ETCS uses special parameters, described in section 12.7, that are used to modify the braking curves in low adhesion conditions. This enables the driver or the ETCS trackside to switch on a low adhesion setting, typically for a certain predefined area. The low adhesion 'slippery rail' setting uses a lower maximum deceleration rate (defined in a national value) to adjust the ETCS braking curves so that the brake warning indication to the driver commences earlier.

ATO over ETCS also includes low adhesion settings that moderate the ATO speed profile to avoid the risk of WSP activation that could otherwise result in a train overrunning platform stopping points.

12.5 ETCS braking curves

Many times every second the ETCS on-board calculates the train braking curves and the supervision limit curves for each target given by the ETCS trackside. A target may be determined by the most restrictive speed profile (MRSP), a stop location (End of Authority, EoA and Supervised Location, SvL) or a Limit of Authority (LoA).

There are three main braking curves used in ETCS. The EBD (emergency brake deceleration) curve is always used, whereas the SBD (service brake deceleration) and driver braking guidance (GUI) curves can be configured and used as required.

The EBD curve specifies the deceleration of the train when the emergency brake is applied. When braking to a stop location, the target is the SvL.

The SBD curve specifies the deceleration when the service brake is fully applied. When braking to a stop location, the target is the EoA.

The purpose of the GUI curve is to provide a comfortable way of braking that reflects good driving practice. It is calculated using a 'normal service brake' model, i.e. a lower deceleration rate.

The supervision limit curves (EBI, SBI1, SBI2, W, P, I) are derived from the train braking curves and are used by the ETCS on-board to tell the driver when to brake, warn the driver in the case of overspeed, and intervene by applying the brakes if the driver continues to overspeed.

If the service brake command interface from the ETCS on-board to the train is implemented and the national value Q_NVSBTSMPERM is set to give permission to use the service brake in target speed monitoring, the first intervention will apply the service brake based on the SBI1 or SBI2 curves. If there is no service brake interface or it is inhibited by the national value, the emergency brake will be applied instead, but the SBI1/2 calculation will use a brake build time set to zero. This will reduce the W, P and I distances to target, which will maximise performance but with an increased risk of an emergency brake intervention.

A full list of braking and supervision limit curves is given in table 12-1, and a typical set for a stop location is shown in figure 12-1.

Curve	Derivation	Use
Emergency brake deceleration curve (EBD)	Based on emergency brake model	Used by the ETCS on-board to predict the emergency braking distance of the train at a given speed
Emergency brake intervention supervision limit (EBI)	Derived from EBD	Trigger for the ETCS on-board to command the emergency brakes
Service brake intervention supervision limit 2 (SBI2)	Derived from EBI	Trigger for the ETCS on-board to command the service brakes (emergency brake if service brake not available)
Service brake deceleration curve (SBD)	Based on full service brake model	Used by the ETCS on-board to predict the full service braking distance of the train at a given speed
Service brake intervention supervision limit 1 (SBI1)	Derived from SBD	Trigger for the ETCS on-board to command the service brakes (emergency brake if service brake not available)
Warning supervision limit (W)	Derived from SBI1, SBI2	Used by the ETCS on-board to warn the driver of an overspeed (via DMI visual indication and audible alert)
Permitted speed supervision limit (P)	Derived from SBI1, SBI2	Used by the ETCS on-board to inform the driver of the permitted speed (via DMI visual indication); typically, the driver will control the train speed just below the permitted speed
Indication speed supervision limit (I)	Derived from P	Used by the ETCS on-board to determine when to start indicating an approaching speed reduction target on the DMI
Guidance curve (GUI)	Optional, modifies the P curve based on a 'normal service brake' model	Used by the ETCS on-board to moderate the P curve (and consequently the I curve) for more comfortable braking
Release speed supervision limit	Calculated by ETCS on-board based on the distance between the EoA (end of authority) and SvL (supervised location) or fixed by ETCS trackside	Used by the ETCS on-board to inform the driver of a release speed (via visual indication)

Table 12-1: Definition of the ETCS braking curves

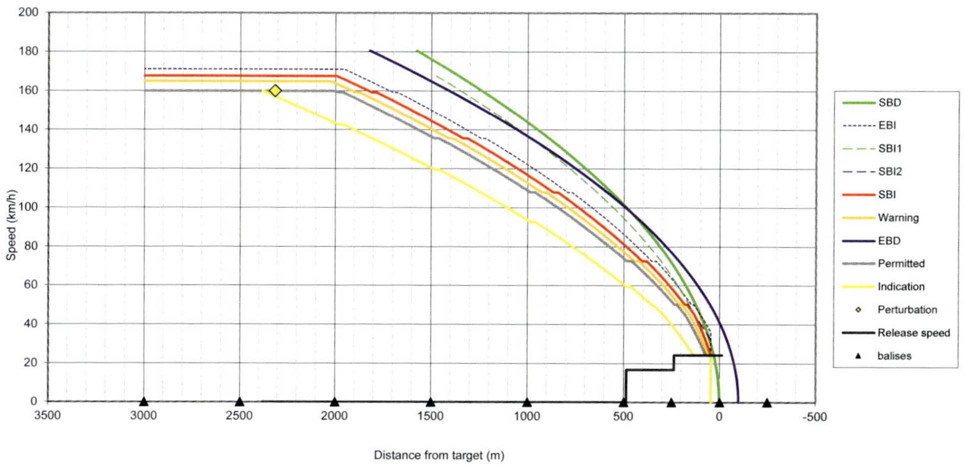

Figure 12-1: Example of ETCS braking curves (ERA braking curves tool)

The braking curves in figure 12-1 have been calculated using a software tool (an Excel spreadsheet) published by the ERA that allows the entry of all the parameters that affect braking and various options for how the brake supervision is to be implemented in the ETCS on-board (such as use of service brake).

The example is based on a passenger train using the 'Gamma model' (described below). It can be seen that the optional service brake intervention has been enabled, the guidance curves have been disabled, and the ETCS on-board calculated release speed has been enabled. On the trackside, balise groups (shown by black triangles on the x-axis) have been spaced at appropriate intervals, and an overlap of 100 m has been defined – this is the distance between the End of Authority (EoA) and the Supervised Location (SvL).

The SBI curve (red) is derived from the most restrictive of SBI1 (derived from SBD) and SBI2 (derived from EBI). In this example SBI2 is the more restrictive and it is not visible because it is hidden behind the SBI curve.

The ETCS on-board calculated release speed is shown by the grey stepped line, and it can be seen that the release speed 'steps up' when the train passes over the repositioning balise group closest to the EoA. This is because the train position confidence interval is reset at the balise, enabling the calculated distance between the maximum safe front end and the SvL to be adjusted (increased), so the train can use a higher release speed and still stop safely at the SvL.

12.6 ETCS braking models

ETCS uses two types of braking model to calculate the EBD braking curve:

The **Gamma model** is applied to trains that have a finite number of pre-defined compositions. For example, a fleet of multiple units may have 4-car, 8-car and 12-car compositions. All the deceleration profiles, correction factors and brake build-up times are pre-configured in the ETCS on-board. At start of mission, the driver will simply select the train type, for example '4-car', and then the ETCS on-board will automatically use the pre-configured data corresponding to that train composition.

The **Lambda model** (also known as the 'conversion model') is applied to trains of variable composition, for example a locomotive-hauled freight train. In this case, it is not practical to pre-program all the brake parameters necessary for calculating the braking curves because there are too many possibilities and combinations. The only practical way to obtain the correct values for the current train composition is to require them to be entered by the driver during the data entry process. Clearly, it is not reasonable to expect the driver to know the deceleration values and brake build-up times. Conversion models are therefore used in the ETCS on-board to convert the parameters entered by the driver (braked weight percentage, brake position, train length, as described below) into the parameters of the corresponding brake model.

Braked weight percentage: This indicates the total braking power of the train. For example, a long freight train may have a braked weight percentage of 60%. By contrast, a loco-hauled passenger train may have a braked weight percentage of 125%. 'Braked weight percentage' is defined in UIC standard 544-1 as the "quotient of braked weight and vehicle mass", and 'braked weight' is defined as the "representative quantity for the mean braking capacity of the vehicle or train, expressed in tonnes". Note that braked weight is actually the total braking force of the train divided by the acceleration due to gravity.

Brake position: Brakes can be selected as 'Passenger in P' or 'Freight in P' or 'Freight in G' as defined in UIC standard 540. This relates to the brake cylinder filling time for air braked trains, which is manually adjustable for some types of freight wagon. For example, the nominal emergency brake reaction time for 'Passenger in P' is 1.42 s, whereas for 'Freight in G' it is 9.30 s.

Train length: This is used by the ETCS on-board to determine the track gradient to be used in the models, to deal with the situation where not all the vehicles are on the same gradient (particularly relevant for long trains). Train length is also a factor in determining the brake build-up time.

One of the main challenges with the Lambda model is how to address the risk that the driver enters the wrong values for braked weight percentage, brake position and train length during start of mission. The ETCS DMI includes a validation process for this data that provides a level of mitigation, but there is still a residual risk that the wrong values are entered. To further mitigate this risk there is the option to limit the choice, for example, setting one braked weight percentage value that applies to all trains, but this results in braking curves that are longer than necessary and a consequential impact on performance.

The logic for use of the Gamma and Lambda models in the braking model calculation is shown in figure 12-2.

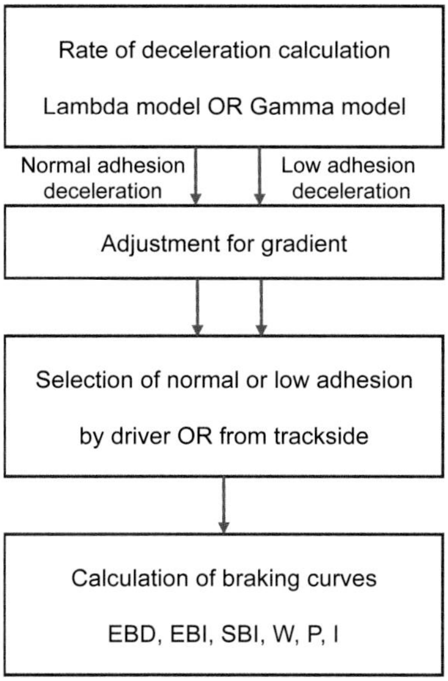

Figure 12-2: Braking model calculation

12.7 Factors used in the braking models

The EBD is expressed as a compound curve, A_safe(V,d). This means that the deceleration 'A_safe' can vary according to both speed 'V' and distance 'd' from the target. The factors involved in calculating the EBD curve differ depending upon the model selected, as shown in table 12-2 below.

Factors used in EBD curve, A_safe(V,d)	Gamma model	Lambda (conversion) model
Gradient	From ETCS trackside	From ETCS trackside
Nominal emergency brake rate (A_brake_emergency)	From train data (stored data originating from specific brake tests undertaken for the train type)	Braked weight percentage entered by driver or via train interface, based on UIC standard 544-1
Train length compensation for acceleration/deceleration due to gravity (gradients)	From train data (stored)	Length entered by driver or via train interface
Rotating masses compensation for acceleration/deceleration due to gravity (gradients)	From train data (stored)	From train data (stored)
Locations with reduced adhesion, based on national values; A_NVMAXREDADH1/2/3 (defining the deceleration rate limits) and Q_NVDRIVER_ADHES (to allow selection by driver)	From ETCS trackside (applied to a certain area) or by driver (switched on/off)	From ETCS trackside (applied to a certain area) or by driver (switch on/off)
Locations without special brake contribution	Track conditions, from ETCS trackside	Track conditions, from ETCS trackside
Correction factors	Kdry_rst from train data (stored data based on failure probability analysis) Kwet _rst from train data (stored data based on specific brake tests undertaken for the train type in wet conditions) M_NVEBCL (Kdry_rst confidence level, acquired from trackside) M_NVAVADH (weighting factor for available rail-wheel adhesion, used to moderate Kwet)	Integrated correction factors (Kv, Kr, Kt) from ETCS trackside Kv relates to the speed and type of train. It corrects for differences between the conversion model and the predicted braking performance for the main types of train. Kr is length dependent, generally longer trains will have better adhesion due to the cleaning effect of the front axles Kt is time dependent and is used to correct for brake build-up time differences that may be influenced by local maintenance practices
Brake position	Not applicable	Passenger in P, Freight in P, Freight in G selected by driver.

Table 12-2: Factors used in EBD calculation for Lambda and Gamma breaking models

Of these factors, the correction factors are the most challenging to determine, in particular the Gamma correction factors, as explained below. Correction factors are used to cater for situations where the train does not achieve its full braking performance because of failures within the train or poor adhesion. National values are used to enable a trade-off to be made between risk, driveability and performance.

Gamma correction factors – braking

The infrastructure manager selects a confidence level for the national value M_NVEBCL to set the minimum required probability that the train's braking system will deliver the stated rate of deceleration when commanded. This confidence level is derived through risk assessment of the consequence of overrunning a supervised location. Table 12-3 shows the probability levels corresponding to the highest and lowest values of M_NVEBCL.

Infrastructure manager's required confidence level	Corresponding M_NVEBCL value	Probability that braking system will achieve [A_brake_emergency (v) * Kdry_rst (v)] when commanded	Impact on performance
High	9	99.9999999% (use lowest set of kdry_rst factors)	Headway worsened
Low	0	50% (use highest set of kdry_rst factors)	Headway improved

Table 12-3: Gamma correction factor for braking

A corresponding derating factor (Kdry_rst) is applied to the nominal emergency brake rate to achieve the required confidence level. Sets of Kdry_rst factors are derived for each type of train by modelling the braking system, taking into account the architecture and the failure probability of the components in the braking system. Kdry_rst factors are needed across the full speed range and for each of the ten confidence levels.

Gamma correction factors – adhesion

Almost all modern trains are fitted with wheel slide protection (WSP). While this improves braking and prevents wheel damage in low adhesion conditions, the braking performance is never as good as on dry rails. For this reason, a set of Kwet_rst factors are derived for each class and consist of train by performing testing on wet rails and at different entry speeds, in accordance with the European standard for WSP (EN 15595). Longer trains tend to have better braking performance in wet conditions due to the cleaning effect of the front wheels. So within the same class of train the Kwet_rst values will vary according to the train length.

Accompanying the Kwet_rst factor the infrastructure manager selects a national value, M_NVAVADH (available adhesion) according to how actively they manage adhesion. Table 12-4 shows the impact of the highest and lowest values of M_NVAVADH.

Infrastructure manager's level of adhesion management	Corresponding M_NVAVADH value	Impact on Kwet_rst factor on the EBD curve A_safe (v,d)	Impact on performance
High	1.0	The Kwet_rst factor is nullified, i.e. no reduction in braking performance in low adhesion.	Headway improved
Low	0.0	The Kwet_rst factor is applied in full.	Headway worsened

Table 12-4: Gamma correction factor for adhesion

12.8 Deriving the supervision limits

When approaching a target, the ETCS on-board runs a series of calculations based on the EBD to derive the supervision limit curves (EBI, SBI1, SBI2, W, P, I). For an EoA the end of the EBD curve is at the SvL (as shown in the previous example). For a MRSP change or LoA the EBD curve is positioned such that it intersects the ceiling speed EBI at the target location.

If the target is an EoA, the ETCS on-board additionally runs a series of calculations based on the SBD. The end of the SBD curve is the EoA. Further calculations are performed for release speed (if set) and the guidance curve (if not inhibited).

If there are multiple targets within the movement authority (for example a MRSP change followed by an EoA), the ETCS on-board performs calculations to derive the supervised limits for each target and then determines the most relevant displayed target (MRDT) in order to display only those limits on the DMI that are relevant to the driver.

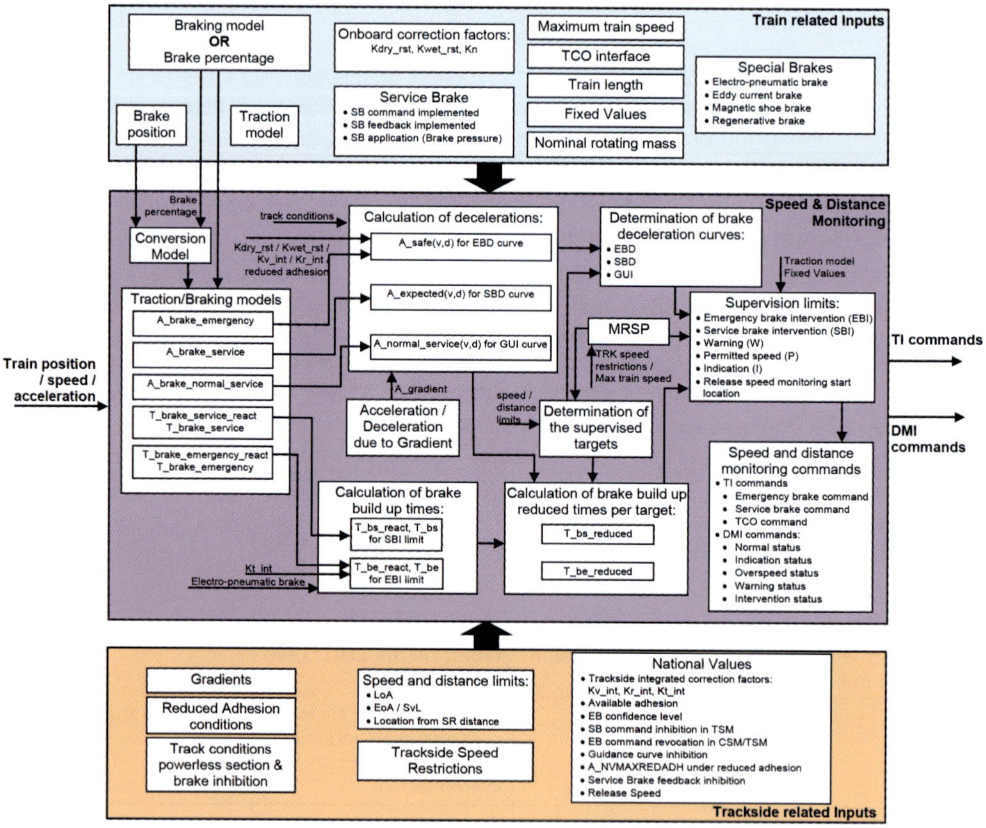

Figure 12-3: ETCS on-board calculation of supervision limits (from SUBSET-026-3)

The supervision limit curves (EBI, SBI1, SBI2, W, P, I) are spaced apart according to driver reaction times, brake build-up times, and other factors such as traction cut-off time. The P-curve is also influenced by the GUI curve (if not inhibited by the national value

Q_NVGUIPERM). It is the I-curve that prompts the driver to start braking. Ideally the driver will brake the train as close as possible to, but not exceeding, the P-curve.

The brake build-up times T_{be} and T_{bs} can be very significant factors, especially for air-braked trains, and there are ways to reduce the impact of these timing factors so that the I-curve moves closer to the target, which is better for performance.

- Include service brake feedback; for example, an ETCS on-board can use a pressure switch to monitor the air pipe pressure and provide positive feedback when the driver is braking the train correctly (following the P-curve) for the target. The ETCS on-board uses this to reduce T_{bs}.
- Set the service brake permitted during target speed monitoring (Q_NVSBTSMPERM) to 'not permitted'. This has the effect of reducing T_{bs} to zero.
- The ETCS on-board can send a command to cut the traction if the train speed V_{est} exceeds the W-curve. This means the traction acceleration (A_{est1}) component can be reduced or even removed because the traction will already have been cut before the train speed exceeds the EBI curve.
- Make use of the new Baseline 4 functionality to calculate reduced brake build-up times (T_bs_reduced and T_be_reduced) if it is determined that a partial brake application is required.

12.9 Impact on headway

The decision to deploy ETCS will almost always be associated with delivering performance improvements (in addition to the safety and interoperability benefits). The performance of the railway system is determined by journey times and headways, which are key ingredients for constructing the timetable (together with other factors such as operational margins, station dwell times, etc).

Journey time is dependent primarily on track and rolling stock characteristics. ETCS has a lesser influence on journey time, for example to enable higher speeds for tilting trains on curves through the use of static speed profile categories. More generally, high-speed operation is only possible with an in-cab signalling system that incorporates automatic train protection (such as ETCS).

Headway, the time interval between two successive trains running at line speed on a section of railway, is dependent upon track, rolling stock and signalling characteristics. For successive trains to run at line speed, the driver of the following train should be able to drive at line speed without seeing any information (whether in-cab or from a lineside signal) that would require braking.

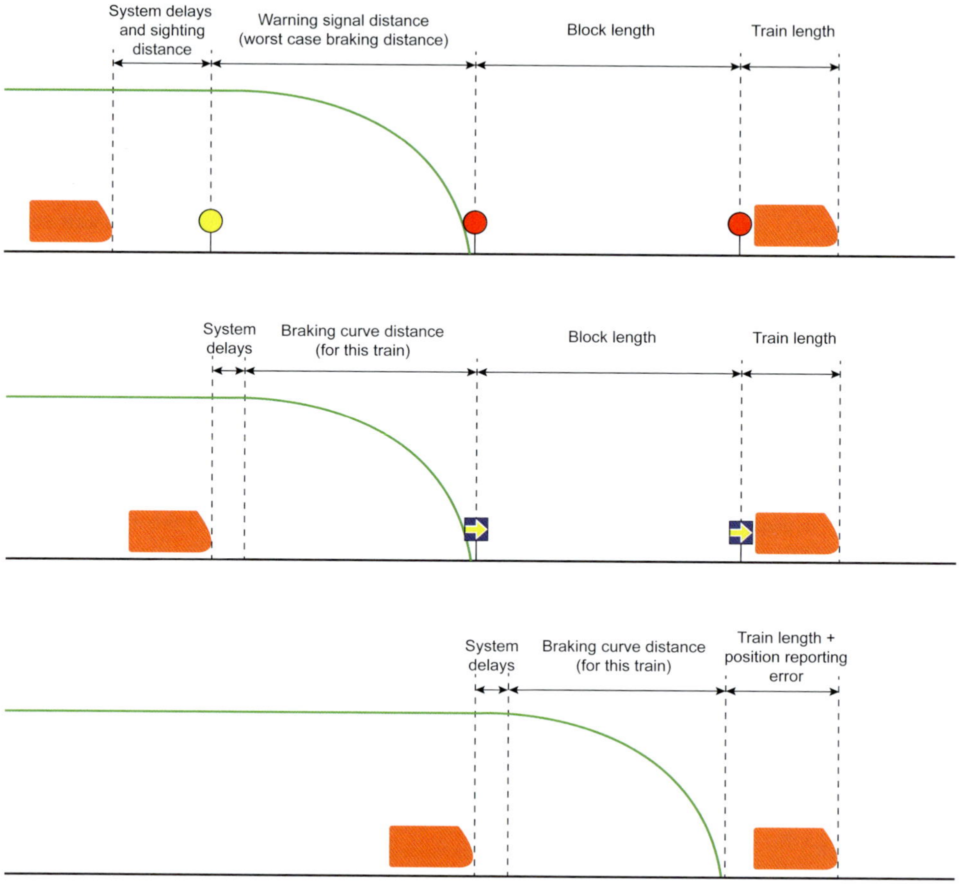

Figure 12-4: Plain line headway for lineside signalling, fixed block and moving block ETCS

With conventional lineside signalling, the dominant factors that determine headway are the distance between successive stop signals (the block length) and the distance from the first warning signal (distant signal) to its corresponding stop signal. This latter distance will be based on a worst-case braking distance for any train expected to run on the line.

For ETCS level 2 without lineside signals, the distance from the first indication to driver of the need to start braking (I curve) to the supervised location becomes the dominant factor. If lineside train detection is used (track circuits or axle counters) the block section length remains relevant, but this can be reduced to much shorter distances than is the case with lineside signals. At low speeds the train length, system delays and signalling block length tend to dominate. At high speeds the braking curve dominates.

If moving block with train position reporting is used (level 2 without lineside train detection), the accuracy and frequency of train position reporting are the main factors in determining headway, in addition to the braking curve. The ETCS on-board train position accuracy degrades with distance from the last balise group used as a location reference, so the spacing between successive balise groups becomes a factor that affects headway. Processing and data transmission times within and between the various components also need to be considered, but these are of secondary significance.

These factors are listed in table 12-5. When calculating the overall headway, the distance factors need to be converted into time by dividing the distance by the line speed.

Lineside signalling with fixed block and lineside train detection	Level 2 no signals with fixed block and lineside train detection	Level 2 no signals with moving block and train position reporting
Distance factors: – Train length – Fixed block length between stop signals (including overlap) – Warning distance from first warning (distant) signal to stop signal – Signal sighting constraints impacting placement of signals	Distance factors: – Train length – Fixed block length (train detection section length) – ETCS braking curve distance from first indication to driver (I curve) to the supervised location	Distance factors: – Train length – Maximum error in train rear position reporting – ETCS braking curve distance from first indication to driver (I curve) to the supervised location
Time factors: – Train detection processing time (track circuit or axle counters) – Interlocking processing time – Signal sighting time for train driver	Time factors: – Train detection processing time (track circuit or axle counters) – Interlocking processing time – RBC processing time – Radio transmission delays	Time factors: – Train position reporting interval – RBC processing time – Radio transmission delays

Table 12-5: Plain line headway calculation factors

The key difference between the ETCS with lineside signalling and ETCS level 2 without signals is that in the latter case ETCS allows the headway to be optimised for each train's speed and braking performance, without the constraint of signal spacing and signal sighting.

For ETCS level 1 or level 2 overlay, the lineside signal spacing tends to be the dominant factor in determining headway. The ETCS braking curves are still relevant, as excessive caution in their calculation could mean that the driver receives an in-cab indication that a speed reduction is required before reaching the sighting point for the warning signal (distant signal). In ETCS level 1 the location of infill balises is also a critical factor, as the time taken for the train to reach the next balise is a variable time delay in the driver receiving advice of an extended movement authority. Care needs to be taken when deploying level 1 on existing lines to ensure these factors do not actually degrade headways.

It should be noted that there are techniques to deal with these lineside signal constraints. For example in ETCS level 2 overlay, the conventional signal blocks can be split into smaller virtual signal blocks, and the conventional signals can be adapted to include a special signal aspect when an ETCS train is approaching. For ETCS level 1 the conventional signals can be replaced by simpler proceed indicators that are less expensive and not subject to the same sighting requirements.

A simple headway calculation is only relevant for one train following another on plain line at the same speed. At a station or a junction where the following train needs to be switched onto a different track, route setting and points movement delays will increase the interlocking processing time, which potentially affects train running times and speeds. Trains that run at different speeds and with differing station stopping patterns also have a major impact, so the true capacity of any railway network needs to take the timetable into account. Translating the plain line headway benefits of ETCS onto usable track capacity may require these other factors to be addressed.

'Real world' performance also depends on the ability of the railway to recover from perturbations. Recovering from a perturbation in level 2 is better because the ETCS on-board can receive an updated movement authority at any location, whereas with conventional lineside signalling or level 1 the train is constrained to the previous movement authority until it passes the next balise.

Another factor that degrades actual headway compared with the calculated value is the train driver response to the lineside signal and DMI indications. Drivers tend to drive more cautiously than the braking curves allow, and there can also be significant differences between individuals. Automatic Train Operation (ATO) enables consistent operation up to the limits permitted by the braking curves, and this is an important enhancement to ETCS capability that is described in the next chapter.

References

SUBSET-026-3: ERTMS/ETCS System Requirements Specification, Chapter 3 Principles

ERA_ERTMS_040026: Introduction to ETCS braking curves

CENELEC EN 15595:2018+A1:2023: Railway applications. Braking. Wheel slide protection

IRS 50540 (formerly UIC leaflet 540) – Brakes – Air brakes for freight trains and passenger trains, International Union of railways, 2024.

IRS 50544-1 (formerly UIC leaflet 544-1) – Brakes – Braking performance, International Union of railways, 2025.

Chapter 13: Automatic Train Operation

13.1 Introduction to ATO

In its most basic form, Automatic Train Operation (ATO) is responsible for controlling a train's movement and braking, instead of the driver performing these functions. ATO is designed to ensure consistent and smooth running, and accurate stopping. When connected to the Traffic Management System (TMS), the ATO can receive real time updates and respond accordingly, to take account of the latest prevailing conditions. For example, to increase speed to achieve a revised station arrival time, or slow down to achieve energy savings. ATO can increase line capacity, improve punctuality and reduce energy consumption.

13.2 Evolution

The world's first ATO railway was London Underground's Victoria Line, opened in 1968. Nowadays ATO has become standard on modern metros and people movers. The Communications-Based Train Control (CBTC) standard for transit systems, IEEE 1474.1, describes ATO as optional, but in reality almost all CBTC systems include ATO.

By contrast, mainline railways have only recently started to adopt ATO. This slow uptake of ATO is attributable to various reasons associated with the complexity of railway operations and the mainline environment, but the growing demand for mainline passenger and freight services in many parts of the world is now driving the need for more efficient rail operations, which ATO helps to facilitate.

'ATO over ETCS' is the term used to describe how ATO is interfaced to an ETCS-equipped train. The first mainline passenger railway to operate with ATO over ETCS was London's Thameslink, in 2018. In this case the need for ATO was driven by capacity. The Thameslink railway has a number of different routes that converge through a central core section, requiring a throughput of 24 trains per hour. Studies showed that this could only be reliably achieved by using ATO.

Around the same time as the Thameslink project, the European Research and Development Programme commissioned a task force to develop the ATO over ETCS specifications for incorporation in the TSI CCS. One of the key objectives was to ensure interoperability such that ETCS/ATO-equipped trains could operate over any ETCS/ATO-equipped tracks. To achieve this, ATO over ETCS uses a standardised air gap interface specification for the transmission of data between track and train during each train journey. This includes both dynamic journey data and static topology data, in a manner similar to ETCS. This is in contrast to CBTC systems that tend to use an on-board map to store all static topology data for the network over which the trains operate. The use of an on-board map is well suited to a self-contained metro system, but is not practicable for a large interoperable railway network.

The new ATO over ETCS functionality was tested on two reference test benches, followed by two pilot projects that were completed in 2020. The first pilot project took place in the UK using a passenger train and involving five different suppliers. This pilot used GSM-R packet switched radio (GPRS) for the transmission of ATO messages. The second pilot project took place in Switzerland and used a freight locomotive, along with LTE (4G) radio for transmission of ATO messages.

Following the success of the pilot projects, the ETCS over ATO specifications were finalised and published in the 2023 version of the TSI CCS, together with updates to the ETCS System

Requirements Specification to include the new Automatic Driving (AD) mode of operation (described in chapter 8).

13.3 Grades of Automation

The various types of ATO are classified in the Grades of Automation (GoA) standard, IEC 62290-1, first published in 2006, as follows:

- GoA2: Semi-automated train operation (STO)
- GoA3: Driverless train operation (DTO)
- GoA4: Unattended train operation (UTO)

GoA2 (STO) still requires a driver in the cab to watch the track ahead of the train, supervise passenger transfer (at stations), engage/disengage ATO, and manage degraded and emergency situations. With GoA3 (DTO), there is no driver but a member of staff is required to be on the train for operating the doors and handling emergency situations. At GoA4 (UTO), all these tasks are performed by the system or by staff at the control centre, i.e. there is no need for a member of staff on-board to undertake these duties.

Although the initial set of ATO over ETCS specifications contained within the TSI CCS are limited to GoA2 (STO), the European Economic Interest Grouping (EEIG) ERTMS Users Group (EUG) has developed an 'ATO Operational Concept' that includes GoA3 and GoA4 for future standardisation.

13.4 Driver Advisory Systems

Driver Advisory Systems (DAS) provide additional information to the driver, to supplement information conveyed by the signalling system. Their purpose is primarily to enable energy savings and capacity improvements.

Standalone DAS (S-DAS) was initially developed for heavy haul freight railways in the USA and Australia. In more recent times it has been deployed on passenger railways in Europe. S-DAS typically uses the Global Navigation Satellite System (GNSS) to track the train's position. It then compares the actual time and location of the train versus a journey profile derived from the pre-defined timetable and issues advice to the driver which aims to maintain punctuality and save energy (speed up/slow down/coast). S-DAS energy saving estimates are typically between 5% and 10%.

Connected DAS (C-DAS) is addtionally interfaced to the traffic management system (TMS) and is able to receive and act upon real-time timetable changes that occur during the day. For example, C-DAS might issue advice to trains to aid a prompt recovery from disruptions, or slow a train in order to avoid it being stopped at a junction because of another late-running train. C-DAS energy savings estimates are typically > 10%, and there may be capacity improvements through reduced junction occupation times.

The ATO over ETCS specifications make provision for C-DAS functionality, but provide flexibility in implementation so that a railway can decide to apply C-DAS but not ATO. The main advantages of this approach are that the C-DAS indications can be integrated into the ETCS DMI (saving physical space) and the train speed/position data is derived from the ETCS avoiding any dependency on GNSS (which will not work in tunnels and may be unreliable in urban environments and in cuttings).

13.5 Architecture and interfaces

ATO over ETCS introduces two new subsystems: ATO on-board (ATO-OB) and ATO trackside (ATO-TS), as shown in figure 13-1. Suppliers can decide whether or not to integrate these new subsystems within their existing products. For example, the ATO-TS can be implemented as an application that runs on the TMS platform. The functionality and interfaces are defined in the following specifications:

- SUBSET-125 is the System Requirements Specification for ATO over ETCS. This describes the ATO functions, driver machine interface, operational states and interfaces.
- SUBSET-126, 130, 139 are detailed Application Layer Interface Specifications between the ATO-OB and the ATO-TS, ETCS on-board and Rolling Stock respectively.
- SUBSET-143, 147, 148 are the corresponding Communication Layer Interface Specifications. Note that the ATO Radio (described in SUBSET-148) can use either GSM-R (Packet Switch), or FRMCS, or a public mobile network operator (MNO).

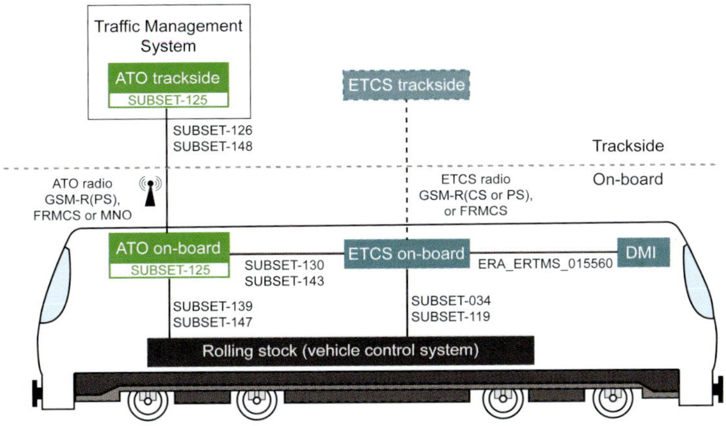

Figure 13-1: ATO architecture and interfaces

The ETCS DMI interface specification, ERA_ERTMS_15560, is shown in figure 13-1 because there are new features that affect the driver, such as ATO data entry windows, ATO symbols (for example, the ATO engage, overshoot/undershoot, door controls/statuses, skip stopping point) and DAS indications (for example, target advice speed, coasting advice).

When running under ATO (ETCS on-board in AD mode), audible overspeed and warning alerts on the DMI are suppressed to allow the ATO to drive close to the ETCS braking curves without causing concern to the driver. For more information about the DMI in AD mode, see chapter 8.

13.6 ATO data flows

The ATO-OB receives a 'journey profile' from the ATO-TS, containing dynamic information about a specific train journey. This includes the times at which the train should arrive and depart specific locations, and intermediate 'timing points', as defined by the TMS.

The ATO-OB also receives 'segment profiles' (typically there are multiple segment profiles associated with one journey profile). These segment profiles contain static information such as gradients, curves, and timing point locations. Segment profiles are stored in the ATO-OB, so they do not need to be repeatedly sent from the ATO-TS for every train journey.

The ATO-OB calculates the optimum speed profile to achieve punctual running in the most energy-efficient manner by combining the information from:

- The journey profile and the segment profiles
- The ETCS supervision limits received via the interface with the ETCS on-board
- The train data

The ATO-OB sends traction and brake commands to the train's control system, and uses the speed and positional feedback obtained via the ETCS on-board to drive the train to the optimal speed profile. This process of calculating the optimal speed profile, generating the traction/brake commands and monitoring the speed and positional feedback, is repeated continually by the ATO-OB to achieve the fine control needed for functions such as accurate station stopping.

Figure 13-2 shows the principal data flows between the ATO and ETCS subsystems, the train traction and braking systems and the driver via the DMI.

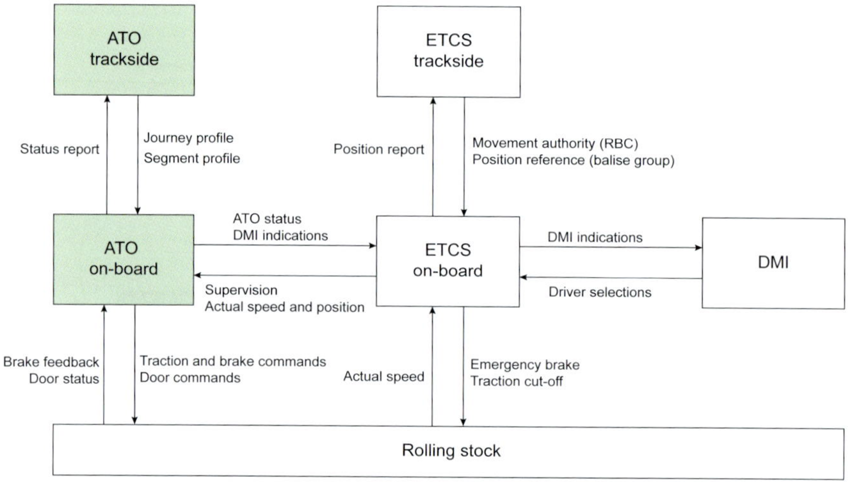

Figure 13-2: ATO data flows (only shows data flows relevant to this description)

13.7 ATO operational states

In a similar manner to ETCS modes, the functionality of the ATO-OB is specified in terms of ATO 'states' and the transitions between them. The states are:

- No Power (NP)
- ATO Configuration (CO)
- ATO Not Available (NA)
- ATO Available (AV)
- ATO Ready (RE)
- ATO Engaged (EG)

Figure 13-3 shows the transitions between these states during a train journey under ATO, starting with configuration of the system after power-on and acquisition of all the data required to allow operation under ATO. The example is based on GoA2 whereby the driver must make a conscious decision to engage ATO after every stop.

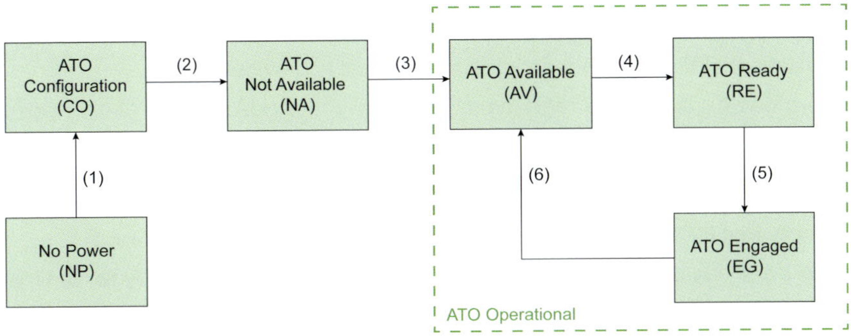

Figure 13-3: ATO-OB states and transitions

Figure 13-4 gives more details of the conditions for the state transitions to take place once the initial configuration is complete (i.e. transitions (3), (4), (5) and (6) in figure 13-3) and the DMI indications for each state.

	Transition (3) > ATO Available (AV), main conditions: ETCS on-board is in Full Supervision (FS) or Automatic Driving (AD) mode and the ATO-OB has a journey profile and the associated segment profiles. The ETCS data and ATO-specific data must be valid. When all conditions are met, the 'ATO selected' indication is displayed on the DMI. If any of these conditions become invalid in AV/RE/EG states, the system reverts to the NA state (transition not shown on figure 13-3).	
	**Transition (4) > ATO Ready (RE)**, main conditions: Direction controller is in the forward position, the ETCS on-board has a movement authority, and the train doors are closed and locked. When all conditions are met, the 'ATO ready for engagement' indication is displayed on the DMI.	
	**Transition (5) > ATO Engaged (EG)**, one additional condition: The driver selects 'ATO ready for engagement' on the DMI. If the ETCS on-board was in FS mode, it will transition to AD mode. When this condition is met, the 'ATO engaged' indication is displayed on the DMI.	
	**Transition (6) > return to ATO Available (AV)**, two alternative conditions: Either the train stops at a stopping location and the train is stationary with brakes applied, or the driver manually applies the brakes using the traction brake controller. When either of these conditions occur, the system reverts to the AV state and the 'ATO selected' indication is displayed on the DMI.	

Figure 13-4: DMI symbols for ATO state transitions

13.8 ATO site engineering

From an engineering perspective ATO is intrinsically linked to both the TMS and the ETCS. Consistency of the data used by these three systems is very important. For this reason, there is a strong focus on creating standardised data formats that can be used as a common source for all topological data for all three systems, together with tools that can convert this data into the particular format needed by each system.

For ATO the primary need is to define the journey profiles and segment profiles for the network, according to the packet formats defined in SUBSET-126 (ATO-OB/ATO-TS Application Layer FFFIS). A brief description of this data is provided below:

Journey profiles contain three main types of data: 'route description', 'timing points', and 'temporary constraints'. This is dynamic information about a specific train journey.

The route description comprises identifiers for one or more segment profiles (described below) and associated version numbers. The version numbers are important because the ATO-OB needs to know that it has the latest valid version of a segment profile.

Each timing point has an identifier. This allows the ATO to associate the dynamic timing point data (time) given in the journey profile with the static timing point data (location) given in the segment profile. The ATO derives the timing point dynamic data from the TMS. For example, a station stop timing point may have an arrival time, a minimum dwell time and a departure time.

Temporary constraints include low adhesion settings, ATO speed restrictions, ATO/DAS inhibition zones, and current consumption limitation zones.

Segment profiles are retained in memory by the ATO-OB, and it will only request a new segment profile from the ATO-TS if the stored segment profile version does not match the one referenced in the journey profile. This might happen if, for example, the infrastructure manager has changed a station stopping point location.

Segment profiles contain static data such as distances, gradients, curves, static speed profile, and traction system information. Importantly, segment profiles also define the timing points in terms of their unique identifier, location, name, stopping tolerance and stopping point reached distance.

Other data contained in the segment profiles include balise group information, platform area definitions, tunnel information, axle load speed profiles, special brake inhibition areas, and unprotected level crossings where the train may be required to stop.

References

IEEE 1474.1-2004: Communications-Based Train Control (CBTC) Performance and Functional Requirements

IEC 62290-1:2025: Railway applications – Urban guided transport management and command/control systems – Part 1: System principles and fundamental concepts

SUBSET-125: ERTMS/ATO System Requirements Specification

SUBSET-126: ERTMS/ATO ATO-OB / ATO-TS FFFIS Application Layer

SUBSET-130: ERTMS/ATO ATO-OB / ETCS-OB FFFIS Application Layer

SUBSET-139: ERTMS/ATO ATO-OB / Rolling Stock FFFIS Application Layer

SUBSET-143: ERTMS/ATO Interface Specification – Communication Layers for On-board Communication

SUBSET-147: ERTMS Data Applications – FFFIS part: CCS Consist Network Communication Layers

SUBSET-148: ERTMS/ATO ATO-OB / ATO-TS FFFIS Transport and Security Layers

ERA_ERTMS_15560: ETCS Driver Machine Interface

Chapter 14: Management of System Versions and Values

14.1 System version management

Since the inception of ETCS, the associated specifications have been constantly evolving to correct errors and ambiguities identified during implementation projects, and to provide additional functionality requested by stakeholders. A new version of the TSI CCS is published every few years, defining the set of documents that make up a new 'baseline' for ETCS. This creates a problem of how to maintain interoperability between new trackside and on-board systems that are compliant with the latest baseline and those that are already in service or under development using an earlier baseline.

To manage this situation, a 'system version' is defined which is incremented with each new baseline. The system version number is in two parts (X.Y), where the first number X distinguishes incompatible versions and the second number Y indicates compatibility within version X. For example, an ETCS trackside system version 2.2 is compatible with an ETCS on-board system 2.1 or 2.3, but not an ETCS on-board system 1.0.

The X component of the version number is incremented for major changes that introduce new functionality to the system as a whole and/or which amend the definition of packets and variables in the ETCS language used for communication between trackside and on-board. The Y component is incremented for error corrections and new functionality that is contained within one of the systems (either on-board or trackside), but which does not require a corresponding change in the other.

A trackside constituent such as an RBC or a balise will always be compliant with a single system version, and this information is embedded in messages from the ETCS trackside to on-board, using the variable M_VERSION. In a level 2 area, the version sent by the RBC determines the operated system version in the whole RBC area, and the version information sent by balises is ignored. This avoids the need to update all the balises when an RBC is upgraded to a new version. In a level 1 area the version sent by each switchable balise determines the operated system version to be used by the ETCS on-board until the next balise is encountered. The 2023 TSI CCS allows a trackside system to be implemented with 7 possible versions – 1.0, 1.1, 2.0, 2.1, 2.2, 2.3, and 3.0.

To achieve interoperability with different versions of trackside subsystems, the ETCS on-board subsystems is required to implement an 'envelope' of multiple system versions. In the 2023 TSI CCS, the permitted options for an ETCS on-board constituent newly placed on the market are 1.0 to 2.1, 1.0 to 2.2 or 1.0 to 3.0. The ETCS on-board adapts its behaviour to use the highest version within its envelope that is compatible with the operated system version that it has received from the trackside. For instance, an ETCS on-board with an envelope 1.0 to 3.0 running in a level 2 area with RBC version 2.2 will operate according to the requirements of version 2.3. A full list of these adaptations is shown in table 14-1.

Trackside system version	System version used by on-board to adapt to trackside		
	On-board envelope 1.0 to 2.1	On-board envelope 1.0 to 2.2	On-board envelope 1.0 to 3.0
1.0	1.1	1.1	1.1
1.1	1.1	1.1	1.1
2.0	2.1	2.2	2.3
2.1	2.1	2.2	2.3
2.2	2.1	2.2	2.3
2.3	2.1	2.2	2.3
3.0	not compatible	not compatible	3.0

Table 14-1: On-board adaption to achieve compatibility to trackside system version

The system version number is not the same as the baseline number. While the documents that comprise a baseline primarily specify the functionality of the latest system version number, they also specify how this should be adapted for the earlier versions. For example, baseline 4.0 generally corresponds to system version 3.0, but within this baseline SRS Chapter 6 specifies how the behaviour of an ETCS on-board that implements an envelope of versions 1.0 to 3.0 is to be adapted for an earlier operating system version that is being used for the ETCS trackside.

This solution gives 'backward compatibility', allowing a railway undertaking (RU) to equip their new trains with the latest versions of ETCS on-board or upgrade existing trains, while retaining compatibility with existing trackside installations. However, there remains a problem when an infrastructure manager (IM) wishes to use the latest version for the further roll-out of ETCS or to upgrade an existing ETCS trackside, and this version is not compatible with an older envelope of versions supported by the ETCS on-board on trains that are already operating in the area. This situation imposes costs on the RUs to upgrade their existing ETCS fitted trains, which can be expensive when obsolescent on-board hardware needs to be replaced to support new software, and the lack of on-board interface standardisation results in the RU having to use the original equipment supplier for the upgrade.

This means that migration between ETCS trackside versions needs to subject to long term planning in a similar manner to the initial migration from national Class B systems to ETCS. The TSI CCS provides a framework for this, placing legal obligations on both IMs and RUs as to how the process will be managed. An example of this is the period of notice required from an IM to a RU for a change in ETCS trackside system version. An additional period of notice applies when a new system version is first defined in the TSI CCS, to allow time for suppliers to develop the new functionality in the ETCS constituents and to place them onto the market.

14.2 Management of ETCS identity numbers

ETCS uses its own language to exchange data between ETCS trackside and ETCS on-board. Within this language is the concept of 'ETCS identity numbers' (NIDs), which appear as variables in the messages to uniquely identify a component of the system that is being referenced, such as a balise group, ETCS on-board or RBC.

To achieve interoperability, the assignment of ETCS identity numbers is managed at two levels:

- NIDs that are assigned centrally by the European Agency for Railways (ERA) as system authority
- NIDs that are assigned by a 'user design authority' for a specific country or region

The relationship between the two levels is managed by the assignment of NIDs by the ERA to each country or region and its associated user design authority. The user design authority is usually an IM or a national standards body. The user design authorities then assign the NIDs that are applicable within the country or region. A user design authority may be responsible for more than one country or region.

The ERA also assigns a unique value for every ETCS on-board, and to identity types of traction system (voltage and frequency) and national train control (Class B) systems. The NIDs assigned by the ERA are:

Value	Meaning
NID_C	Identity of a country or region
NID_XUSER	Identity of user design authority
NID_ENGINE	On-board ETCS identity
NID_NTC	National system identity (Class B train control system)
NID_CTRACTION	Identity of the traction system

Table 14-2: NID values assigned by the ERA

The procedure to request the assignment a unique value from the ERA is described in the document ERA_ERTMS_040001 "Assignment of Values to ETCS variables", which is regularly re-issued with lists of all the values that have been assigned. A request can be submitted by a RU, an IM or another stakeholder such as a rolling stock leasing company or a supplier. For the on-board identity (NID_ENGINE) the ERA will usually issue a range of values to be used for the ETCS on-boards to be installed on a specific fleet (or type) of trains. Although this process is only mandatory within the European Union, the ERA is willing to assign values for ETCS projects anywhere in the world.

The NIDs that are assigned by a user design authority apply to elements of the ETCS trackside subsystem, and they are unique within a region/country defined by one value of NID_C. The combination of NID_C and the specific NID (assigned by the user design authority) then uniquely identifies the element globally. Identities of this type are:

Value	Meaning
NID_BG	Identity number of a balise group
NID_RBC	Identity number of a radio block centre
NID_MN	Identity number of the GSM-R network
NID_RADIO	Radio subscriber number, i.e. the 'telephone number' of the ETCS entity (RBC or on-board)
NID_KMC	Identity number of a key management centre

Table 14-3: NID values assigned by user design authorities

14.3 Management of national values

Within ETCS there are a number of parameters, called 'national values', that can be configured to best suit each specific railway operating environment. Even within a single national environment there may be regions where different national values are required, for instance to optimise performance or provide enhanced levels of safety. Each region is allocated a NID_C, and each NID_C has its own set of national values.

As with the identity numbers, the assignment and control of national values need to be controlled by the user design authority for the country or region. The national values are transmitted to the ETCS on-board via packet 3, either as part of a balise telegram or a radio message, when a train approaches an area for which a different set of national values apply. The ETCS on-board then uses this set of values until it receives a further update.

In baseline 4 there are 36 national values that are defined, although not all of them are necessarily used in every specific region or country. There are five types of national values, relating to speed, distance, time, miscellaneous and qualifiers. Table 14-4 provides some examples of each type.

Type	Example	Meaning of example
Speed	V_NVSTFF	SR mode speed limit
Distance	D_NVSTFF	SR mode maximum permitted distance
Time	T_NVCONTACT	Safe radio connection timeout
Miscellaneous	M_ NVCONTACT	Reaction to be performed when T_NVCONTACT timer elapses
Qualifiers	Q_NVLOCACC	Default location accuracy of a balise group

Table 14-4: Examples of national values of each type

References

SUBSET-026-6: ERTMS/ETCS System Requirements Specification, Chapter 6 Management of older System Versions

ERA_ERTMS_040001: Assignment of Values to ETCS variables

Part 3
ETCS Subsystems and Interfaces

In Part 3, the constituent parts of ETCS on-board of the train, on the track and in the control centre are described. The key interoperable interfaces between track and train via balise and via radio are explained, together with the means by which train location is tracked. The interfaces with other railway systems and the train driver are described. The final chapter explains the certification process which assures the interoperability of ETCS constituents from different suppliers.

Chapter 15: Trackside Boards

15.1 Requirement for trackside boards

In conventional lineside signalling, trackside signals and signs convey information to the train driver that is linked to their specific location. ETCS provides the equivalent information via the DMI in the train cab, but in some circumstances the DMI information needs to be supplemented by trackside signs.

Where ETCS is implemented as an overlay on conventional lineside signalling, the signs associated with the signalling are retained, but for ETCS level 2 without lineside signals a standardised set of signs has been specified in the CENELEC standard EN 16494:2025 Railway Applications – Requirements for ERTMS Trackside Boards. It should be noted that this standard and other ETCS documents refer to 'boards' and 'marker boards' rather than 'signs', but the meaning is the same.

There are two main purposes for providing these trackside marker boards:

- To provide clearly defined position references that enable train drivers to unambiguously understand their location when ETCS is operating in a degraded mode.
- To provide precise 'real world' visual references to confirm the exact locations of transitions and track conditions that are indicated on the DMI.

EN 16494:2025 defines the design and optical properties of the boards. The shapes and proportions of the boards and the symbols and lettering on them are specified, but the overall size is not mandated. The standard gives examples for 500 mm and 700 mm high boards but allows for smaller boards to be used in areas of low speed and limited clearances, and larger boards on high-speed lines.

Guidance on the purpose and use of marker boards is provided in ERA Technical Document 011REC1053 "Harmonised ERTMS Marker Boards", but the details of where and how they are to be used, and how their use is linked to the operating principles adopted for ETCS level 2 by a national infrastructure manager, are not harmonised by the TSI CCS.

15.2 ETCS Stop Marker

The most widely used of the ETCS defined marker boards is the stop marker. This is a distinctive yellow arrow on a blue background. The marker board is mounted on a structure to the left or right of, or above, the track to which it applies, with the arrow pointing towards the relevant track, to avoid any possible confusion in multiple track areas.

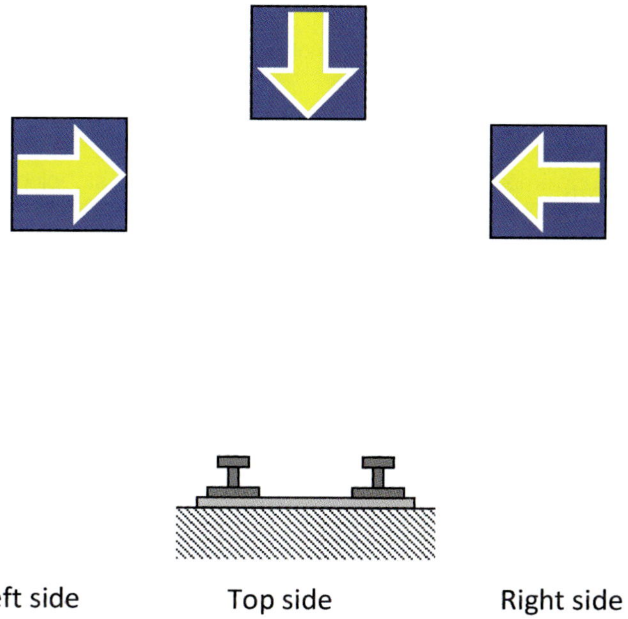

Figure 15-1: Installation options for ETCS stop marker (from ERA Technical Document 011REC1053)

Each ETCS stop marker has an associated clearly visible and unambiguous identifier (letters and/or numbers) that can be read by the train driver from the cab of a train when it arrives at the marker. The format and content of the identifier is not harmonised; it will typically follow the national rules for signal identifiers – see figures 15-3 and 15-4.

The ERA guidance on the application of the ETCS stop marker is that it:

- shall be used to unambiguously identify an EOA (end of authority) which protects one or a group of safety-critical points such as a switch, a conflicting route as entrance of a station, junctions, etc. where the driver requires supplementary information to identify the exact stopping location in addition to the display on the DMI (for instance where a release speed is provided).
- may be used to unambiguously identify the end of a block section as an EOA where the driver requires supplementary information to identify the exact stopping location in addition to the display on the DMI (for instance where a release speed is provided).
- may be used to identify other points of the track the driver shall not overpass when a movement authority (MA) is not available, unless he has received a specific authorisation by the signaller.
- serves to mark the location up to which the driver needs to confirm that the line is clear (e.g. for Track Ahead Free (TAF) functionality) or he has to drive under his responsibility before a possible MA in FS/OS mode is received.

15.3 ETCS Location Marker

The ETCS location maker has the same general format as the ETCS stop marker, but with a black arrow on a white background.

The ERA guidance on the application of the ETCS location marker is that it:

- may be used to unambiguously identify the end of a block section as an EOA where the driver requires supplementary information to identify the exact stopping location in addition to the display on the DMI (for instance where a release speed is provided).
- may be used to identify other points of the track the driver can overpass without a specific authorisation by the signaller when an MA is not available.
- serves to mark the location up to which the driver needs to confirm that the line is clear (e.g. for TAF functionality) or they have to drive under his responsibility before a possible MA in FS/OS mode is received.

Many ETCS applications only use the ETCS stop marker. The ETCS location marker has been introduced to allow a distinction between locations the driver must not pass when an MA is not available, unless they have received a specific authorisation by the signaller (stop marker), and other locations that the driver can pass on their own authority (location marker).

15.4 Track Condition Markers

The CENELEC standard also defines a selection of marker boards to provide a precise visual reference for locations where transitions occur, or track conditions apply. These include:

- ETCS level transition
- GSM-R network boundary
- Raise/lower pantograph
- Electrification neutral zone
- Change of electrification traction system (AC/DC, voltage, none)
- Sound warning horn
- Safe stopping and non-stopping areas
- Inhibit special brakes (magnetic shoe, eddy current, regenerative)
- Level crossing

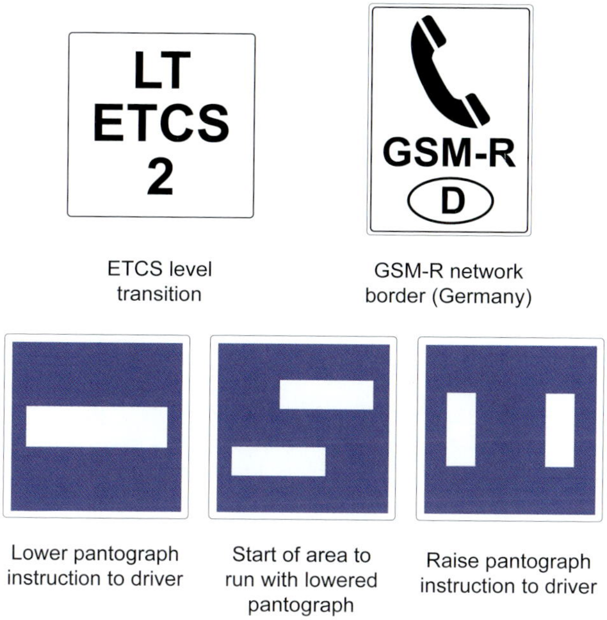

Figure 15-2: Examples of ERTMS track condition marker boards

15.5 Other trackside signs

Other trackside signs in ETCS level 2 areas are not harmonised by international standards, so they will generally follow the signing conventions set out in existing national standards. Such signs may be required to support degraded modes of operation when the necessary information cannot be displayed on the DMI. For example, in Staff Responsible (SR) mode the DMI only displays the national value for maximum speed in this mode, so signs may be needed to inform the driver where speed limits need to be more restrictive than the national value.

Figure 15-3 shows an example of an ETCS stop marker in the format specified in EN 16494, with an identifier and speed sign that are in accordance with national standards.

Figure 15-3: Example of ETCS stop marker with national identifier and speed signs (Cambrian line in Great Britain)

National signs may also be required on lines that are dual fitted with ETCS and a national system. Figure 15-4 shows an example where an ETCS stop marker is installed alongside a stop marker for the national system.

Figure 15-4: Combined stop marker for ETCS and national system TVM (LGV Est line in France)

References

CENELEC EN 16494:2025: Railway applications – Requirements for ERTMS Trackside Boards

ERA Technical Document 011REC1053: Harmonised ERTMS Market Boards

Chapter 16: Eurobalise Track to Train Data Interface

16.1 Eurobalise Transmission System

The Eurobalise Transmission System is used for conveying safety-related information from the trackside to the train. The system comprises the trackside-mounted Eurobalise and the on-board transmission equipment. It is a spot transmission system, meaning that the transfer of data occurs only in the vicinity of each balise. The on-board equipment has two main components, an Antenna Unit and a Balise Transmission Module (BTM). In terms of the Eurobalise there are two main types; fixed (transmits fixed data to the train, with no connection to other trackside equipment) and controlled (transmits variable data, with a cabled connection to other trackside equipment). Eurobalises are available in two sizes; standard and reduced.

The on-board transmission equipment continuously generates a signal which activates the track-mounted balises as it passes over them, a process called tele-powering. The vehicle-mounted antenna unit transmits this signal to the balise via the airgap. When received by the balise, the signal provides power to activate it, which initiates the transmission of data to the train.

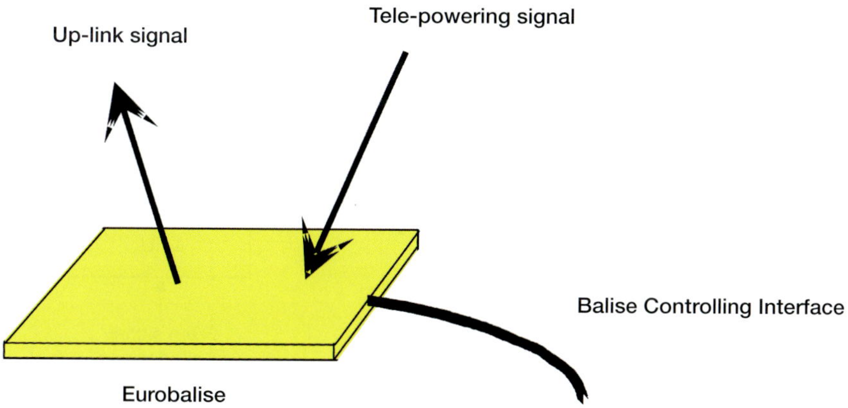

Figure 16-1: Eurobalise and its interfaces

There are four top-level functions defined for the Eurobalise Transmission System. These are as follows:
- Balise detection
- Transmission of safety-critical data from trackside to on-board
- Provide data used for accurate determination of the train location
- Allow understanding of the direction of travel of the train. The ETCS on-board system evaluates the driving direction on the basis of the reported sequence of the Eurobalises passed.

Data is transmitted to the train only when the antenna unit passes over, or is standing over, a Eurobalise. The length of track over which the information is passed is limited to approximately 1 m per Eurobalise.

The Eurobalise transmission system is used in both ETCS level 1 and 2, and for transitions to and from level 0 and level NTC. Level 1 uses controlled Eurobalises, interfaced to the lineside signalling system by a Lineside Electronic Unit (LEU) so that variable data generated by the signalling can be transmitted to the train. Level 2 uses fixed Eurobalises which always send the same information, either a simple location reference or static information such as gradients. In level 2, variable signalling information is sent via radio from an RBC.

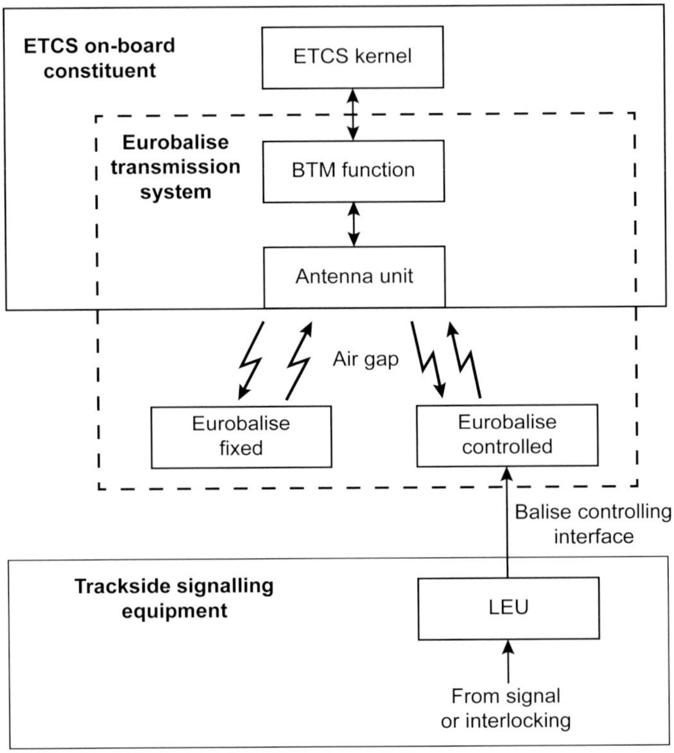

Figure 16-2: Eurobalise Transmission System

In general, the Eurobalise and air gap specifications are very rigid whereas the specification of the on-board transmission equipment is more flexible. The balise transmission module (BTM) is specified in terms of the functionality of the ETCS on-board equipment, rather than as a separate component. This leads to some differences between ETCS suppliers' on-board products in terms of packaging and integration. In this chapter we use the term 'ETCS kernel' for the main part of the ETCS on-board with which the BTM interfaces.

In the functional interface between the BTM and the ETCS kernel, the following information is exchanged:

From BTM function to Kernel:

- Decoded Telegram (derived from the data transmitted by each balise)
- Localisation Data
- Error Reports
- Operational Mode Status

- Remote powering Status
- Health Status (including start-up tests)

From Kernel to BTM function:

- Time and Odometer Data
- Orientation of the train (front/rear)
- Operational Mode command
- Remote powering on/off command

16.2 Eurobalise

The track-mounted Eurobalise is interrogated and powered by means of an inductively coupled remote powering signal. The response of the Eurobalise is also an inductively coupled up-link signal to the train passing over it. The vehicle-mounted antenna unit is both the transmitter of the power signal and the receiver of the up-link signal. The data carried by the up-link signal is held in a non-volatile memory in the Eurobalise or, in the case of level 1 applications, is received by the Eurobalise via a serial data link from a LEU. The data in the non-volatile memory is programmed when the Eurobalise is installed on the track. In level 1 applications using controlled balises, if the interface with the LEU is disturbed or interrupted the Eurobalise automatically switches over to transmitting the telegram in its internal non-volatile memory.

Upon activation by the powering signal from a passing train, the Eurobalise responds by transmitting a 4.2 MHz frequency shift keyed up-link signal.

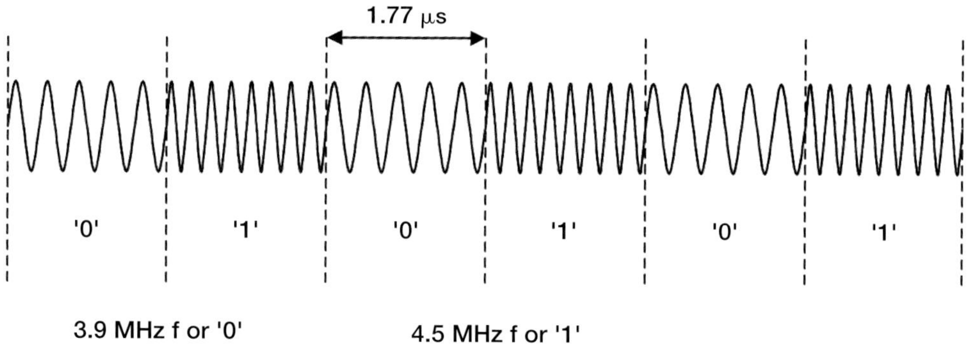

Figure 16-3: FSK up-link signal for Eurobalise

Data transmitted from track to train is considered to be safety-critical. Protection of this data against air-gap noise effects and noise-induced hazards in both the receiving and transmitting equipment is sufficient to ensure error detection to the extent necessary for safety (to comply with specific coding requirements). The coding features a process whereby unprotected data (so-called User Bits) is scrambled according to defined algorithms, thus forming a specific telegram format. This has excellent protection against problems such as random bit errors, burst errors, bit slips, and bit insertions. The telegram format allows quantitative evaluation of the effect of random interference.

The key features of the telegram format are the following:

- Two compatible telegram lengths, 1023 and 341 bits respectively.
- From these telegram lengths, 830 and 210 bits respectively are used to transmit information.
- Protection against various types of transmission errors that could result in unsafe data being used by the onboard ETCS.
- Inversion of any bit of the telegram is always recognised by the decoder.
- Transmission need not start (or end) at the beginning of a telegram.

Figure 16-4: Reduced size controlled Eurobalise with cable to LEU

The specifications for the standard size and the reduced size Eurobalises define a physical output field strength and reference area over which an adequate field strength must exist for the purposes of transmitting information to trains. The standard size Eurobalise has an active reference area of 358 mm by 488 mm, and the reduced size Eurobalise has an active reference area of 200 mm by 390 mm. The physical size of the actual Eurobalise is somewhat larger because of the need for encapsulation, and the thickness is a few centimetres. The thickness of the balise encapsulation is not defined in the standard, but the distance between top-of-rail and the antenna loop inside the balise is defined.

Eurobalises are used in a large variety of environmental conditions such as water, salt water, snow, ice, iron ore, etc. To provide some flexibility, they are divided into two different classes, A and B. Class A Eurobalises are intended to be used in a more demanding environment than Class B Eurobalises. For both, standard size and reduced size balises are possible.

Eurobalises are mounted in the middle of the track, with an equal distance to both rails within a certain tolerance, and within an overall height range of 93 mm to 210 mm below the top of the rail. In practice each combination of balise size and environmental class may further constrain the height range within which a balise can be mounted. The height range may also be limited if the balise is installed near metallic structures.

There must be a minimum distance between consecutive balises along the track. The minimum distance depends on the size of the balise and the maximum train speed at which it needs to be read. Below 180 km/h the minimum spacing is 2.6 m for standard size Eurobalises and 2.3 m for reduced size Eurobalises. For both types, the minimum spacing is 3.0 m for 180–300 km/h and 5.0 m for 300–500 km/h.

Crosstalk can occur where a BTM receives data from a balise that is not intended to be read, for example a balise on an adjacent track. The Eurobalise Transmission System protects against crosstalk by evaluation of received signal levels, but this depends on the lateral spacing between the BTM (the centreline of the track that the train is running on) and the balise that is generating the crosstalk. A minimum spacing of 3.0 m is required to prevent crosstalk from a balise that could be transmitting in response to a tele-powering signal from a train on a parallel track. A reduced lateral spacing of 1.4 m is allowable if the crosstalk balise is on a converging track which means it cannot receive a tele-powering signal from another train. There is also a minimum vertical spacing requirement to cater for the scenario where a balise is installed on a track under a bridge below another railway line.

In many cases two or more Eurobalises are installed at the minimum spacing along the track to form a 'Balise Group'. The sequence of balises within the group encountered by a train allows the direction of movement to be deduced by the ETCS on-board. A balise group might also be required if the amount of information to be sent to the train exceeds the capacity of a single balise. In level 1 areas, usually one of the balises in the group will be a controlled balise, sending variable data. If a second balise is required to allow the direction of travel to be determined, this can be a fixed balise.

In most cases balises are permanently installed in the track, but temporary installation is a possibility, for example to provide temporary speed restriction information in ETCS level 1.

16.3 Lineside Electronic Unit

In ETCS level 1, controlled Eurobalises are usually installed at each signal and they provide movement authority information equivalent to the signal aspect. The data translation is undertaken in a lineside electronic unit (LEU) located in an equipment cabinet beside the signal. Various types of electrical connection between the signalling equipment and the LEU are used, either directly to the signal, or via relays or to an object controller linked to an electronic interlocking. Where ETCS is being installed alongside or instead of a national train control system with similar functionality, the same type of interface may be used. The ETCS standards only define the interface from the LEU to the Eurobalise, so each ETCS supplier can produce their own design of LEU, with variants to interface with different types of lineside equipment.

The LEU continuously transmits a telegram to the Eurobalise. However, the Eurobalise starts sending the telegram only when it is energised by the powering signal from a passing train, even if that happens in the middle of a telegram from the LEU. When activated by a train, the Eurobalise optionally sends a message to the LEU which prevents any changes to its telegram content for a minimum of 10 milliseconds. This enhances availability, so that at least one complete telegram is sent to the train even at high speed and in noisy environmental conditions. If a Eurobalise cannot detect any valid signal or valid data at the LEU interface, it sends a stored (default) telegram. This applies both to the situation when the Eurobalise has been activated by a passing train and is ready to start sending a telegram, and also to the situation when the Eurobalise can no longer detect any valid signal on the LEU interface during the passage of a train.

One of the drawbacks of ETCS level 1 is that information on-board the train regarding the state of the signalling system is only updated when the train passes a balise. If balises are located only at signals, then whenever a signal aspect changes a discrepancy will exist between the on-board supervision of the train and the next lineside signal (which may or may not be visible to the driver) until the train passes the signal. Trains must slow down (as determined by the on-board supervision) even when a signal has cleared in front of them, which limits performance. To reduce this negative impact, additional 'infill' balises may be provided, to update the on-board information about the state of the signalling ahead before reaching the next signal. The infill information can be sent from a balise already provided for another purpose, such as one at a signal for trains running in the opposite direction. Alternatively, if an additional balise has to be provided for infill purposes only, this can usually be a single balise, connected to the LEU at the signal ahead.

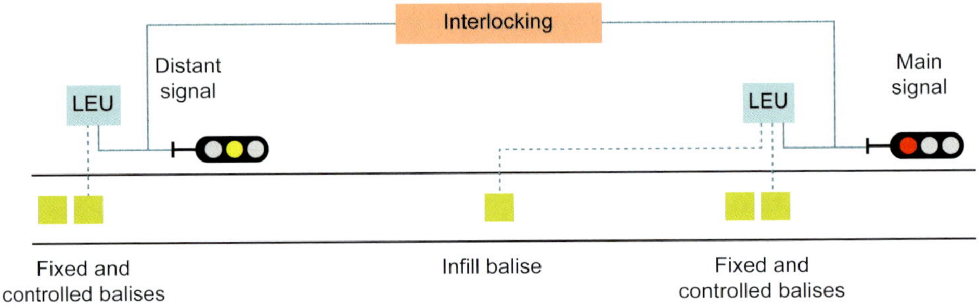

Figure 16-5: Eurobalise installation for level 1 with lineside LEUs

An alternative architecture for level 1 is to provide a direct connection from the interlocking to each controlled Eurobalise. This is sometimes referred to as a 'centralised LEU' or a 'balise driver module' in the interlocking. This reduces the need for additional hardware and wiring in the lineside equipment cabinet at the signal, and also allows greater flexibility in the type of information sent via the balise to the ETCS on-board system. For example, longer movement authorities can be provided using information that is known in the interlocking for multiple signals ahead of the train, and additional functions such as dynamically updated temporary speed restrictions can be implemented.

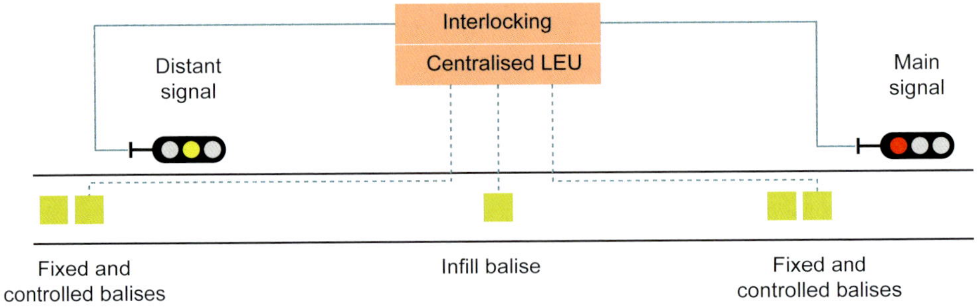

Figure 16-6: Eurobalise installation with Centralised LEU architecture

While LEUs are mostly used in level 1 applications, they can sometimes be useful in level 2. This applies especially to locations where transitions between levels occur, and at RBC-to-RBC handover locations. For instance, where there are diverging routes in a transition area the announcement of level transition can be provided only to those trains routed into the ETCS level 2 area.

16.4 On-board balise transmission equipment

The on-board balise transmission equipment is part of the ETCS on-board system. Its main functions are:

- to generate tele-powering signals to the balises,
- to receive and process up-link signals from the balises,
- to provide the interface between the airgap and the ETCS kernel, and
- to provide applicable test data to the Test Interfaces via an Interface Adapter.

The on-board transmission equipment provides the tele-powering signal used for activating balises. For the basic system, the tele-powering signal is a 27 MHz Continuous Wave (CW), used only for activating Eurobalises.

Upon activation, the Eurobalise responds by transmitting a 4.2 MHz frequency shift keyed up-link signal containing the coded data. On receipt of the data, the BTM unscrambles and checks it to determine if there are any errors and then forwards the data along with any error code to the ETCS kernel. The ETCS kernel applies a safe reaction dependant on the error code.

Accurate train location is achieved by the BTM reading the balise multiple times as it passes over it and then comparing the Received Signal Strength Indication (RSSI) for each reading. The read with the highest RSSI value corresponds with the point at which the antenna unit is closest to the mid-point of the balise. The BTM converts this reading into an accurate location by combining the balise information with the speed and distance information from the on-board odometry system. The BTM then sends the accurate location data to the ETCS kernel. The location accuracy for safety purposes is to within ±1 m with respect to the centre of the Eurobalise. Significantly better location accuracy is achievable for non-safety purposes, an example of which is station stopping accuracy for Automatic Train Operation.

In addition to passing protected data and accurate location data to the ETCS Kernel, the BTM function also reports balise detection. This happens even when a Eurobalise is detected but no telegram is decoded. Although failure to decode a telegram is not safety-critical, it is a safety-critical failure not to detect a balise.

ETCS allows for the fact that the balise telegrams may change under the passing train, for example if the signal aspect changes. A message counter is included in the balise telegram header for this purpose.

Some older national balise transmission systems (most often referred to as the KER systems – KVB, EBICab, and RSDD) use remote powering similar to Eurobalise. However, for these systems the tele-powering signal is a 27 MHz modulated signal where all modulation pulses have the same length. This is referred to as 'non-toggling modulation'. Eurobalises are designed not to respond to this non-toggling modulation, which prevents KER-only trains being 'disturbed' by responses from Eurobalises (the worst case being an emergency brake application).

An optional function for the ETCS on-board system makes it possible for trains that are fitted with both ETCS and a KER to activate and receive information from both Eurobalises and KER balises using the same antenna unit. In this case, the tele-powering signal is a 27 MHz modulated signal using 'toggling modulation'. Toggling means that the pulse length of the modulation differs between two successive pulses.

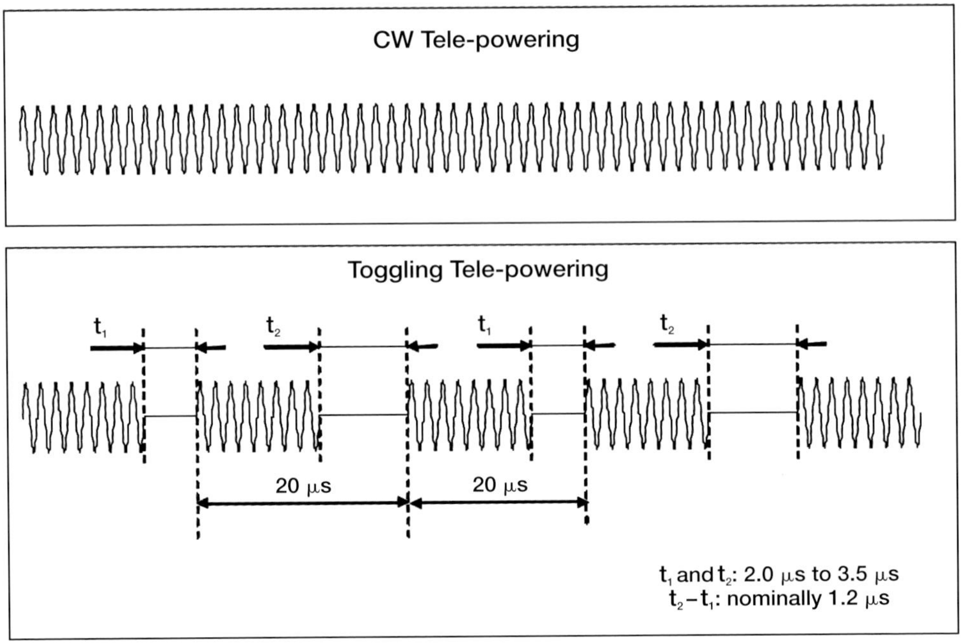

Figure 16-7: Tele-powering signal (CW and Toggling)

Table 16-1 summarises the behaviour of Eurobalises and KER balises to the different types of tele-powering signal:

Tele-powering signal	27 MHz Continuous Wave	27 MHz Toggling Modulation	27 MHz Non-Toggling Modulation
Application	ETCS on-board (mandatory)	ETCS on-board (optional)	KER on-board (legacy)
Eurobalise	Responds	Responds	No Response
KER Balise	No Response	Responds	Responds

Table 16-1: Eurobalise and KER balise response to tele-powering signals

16.5 Semi-continuous infill using Euroloop or radio

As has already been said, a Eurobalise used as a spot infill device can only update the on-board information at the balise location. The potential for a discrepancy between the information held by the on-board system and the aspect shown by a signal ahead takes place from when a signal aspect changes until the train passes the balise sending the infill information, and again if the signal aspect changes when the train is between the infill balise and the signal.

In the early development of ETCS this was a significant concern for railways using an existing national train control system which featured coded track circuit or inductive loop transmission to trains, as these systems provide continuously updated information on approach to a signal. In other words, the national systems were better than ETCS in respect of this problem. To address the problem, the ETCS standards included two optional enhancements of ETCS level 1, to provide semi-continuous infill. These were 'Euroloop' (a leaky feeder cable running along the rail) and 'Radio Infill' using GSM-R. However, in practice neither has been widely adopted and the requirement to provide the optional feature on-board is a barrier to interoperability. Projects where spot infill is considered inadequate have instead adopted ETCS level 2 which inherently provides continuous infill. From 2020 onwards new installations of Euroloop or Radio Infill were no longer permitted and so they are not described any further in this book.

16.6 Virtual Balise Cover

The Virtual Balise Cover (VBC) is a new feature that was introduced as part of ERTMS Baseline 3. The VBC allows the ETCS on-board system to ignore specific Eurobalises and is designed to be applied and removed quickly and efficiently. This is very useful when migrating an existing line to ETCS because it means that new balises can be installed progressively, e.g. over several months, without affecting the ETCS equipped trains already using the route.

The prerequisite for the VBC feature is that all Eurobalises need to be programmed with a VBC marker, a simple identity number. Typically, the same VBC marker will be applied to all Eurobalises within a certain commissioning area.

VBCs can be set or removed by the driver (entering VBC data into the DMI) or via trackside equipment, typically by positioning temporary Eurobalises at the boundaries of each commissioning area. The data includes the VBC Marker, the Country/Region Identifier and the Validity Period. Once set, the ETCS on-board system will ignore any Eurobalise that contains the VBC identity (a combination of a VBC marker and the country/region identity) and is within the VBC validity period.

References

SUBSET-036: FFFIS for Eurobalise

Chapter 17: Train Location in ETCS

17.1 Location reference at a balise

A fundamental requirement for safe operation is that each train has an accurate knowledge of its current location, speed and direction of movement that can be matched to the movement authority and other information (gradients, speed profile, etc.) received from the trackside systems.

It is important to understand that the concept of 'location' in ETCS is not a position on a geographical map in two or three dimensions as with a GNSS satellite positioning system. Instead, all location information on-board is registered as a distance and direction in one dimension along a railway track from a location reference balise. This avoids any requirement for the ETCS on-board to store or access a 'map' of the railway network.

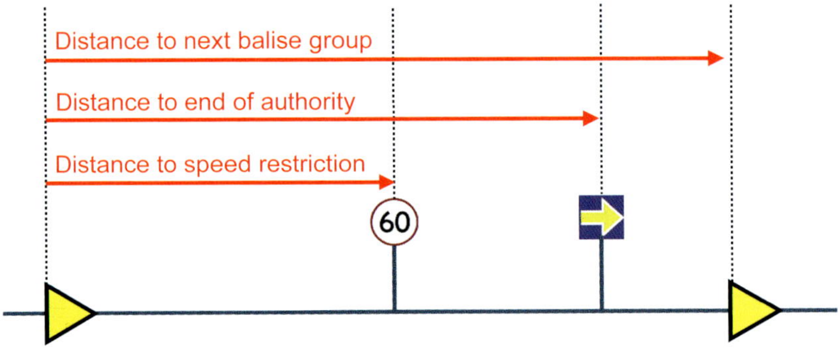

Figure 17-1: Examples of position referencing from a balise group

In level 1 the distance and direction information is always given with reference to the balise group giving the information, while in level 2 the radio messages include a reference to the unique identity of a specific balise group in the variable NID_LRBG (last relevant balise group). The ETCS on-board maintains a record of up to 8 balise groups that have been passed and the distance travelled from each of them, to ensure the distance information received from the trackside can be matched to the current location of the train in the event that NID_LRBG does not refer to the most recently passed balise group.

The maximum signal strength received by the on-board BTM when passing over a balise is used as the point for registering the precise train location, and the order in which the individual balises are read within a balise group defines the direction of movement. Balise groups usually consist of two or more balises and the location reference of the balise group is the location of balise number 1 in the group, regardless of the direction in which the balise group is passed. The balise number is given by the parameter N_PIG in the header of the balise telegram. Distances from the location reference are defined with respect to the orientation of the co-ordinate system of the balise group. The co-ordinate system ensures common location referencing between onboard and trackside systems for all location-based information that is exchanged. The nominal (normal) direction of the co-ordinate system corresponds to balises passed in increasing number sequence; reverse direction corresponds to balises passed in reducing number sequence. A single balise can be used for location reference if the direction of passing the balise can be deduced from linking information referenced to a previous balise group.

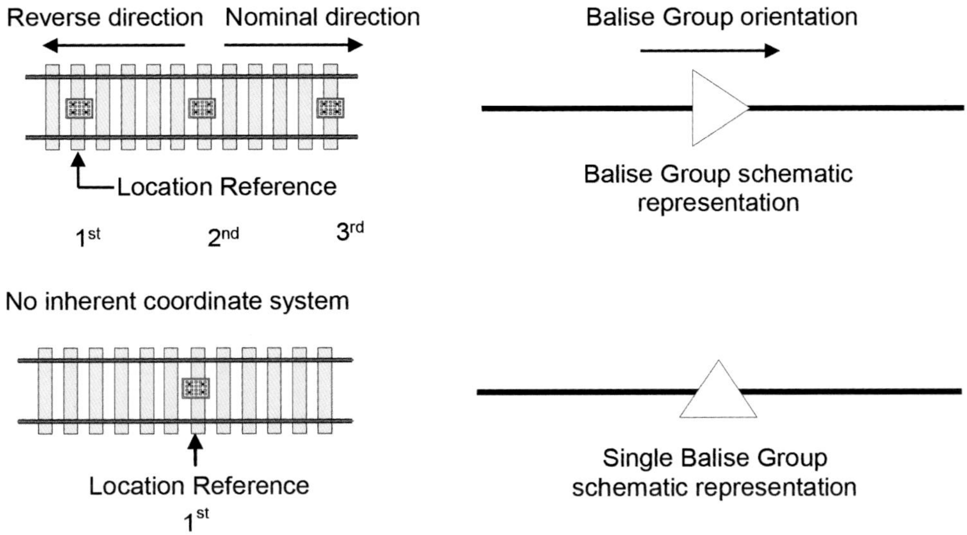

Figure 17-2: Orientation of balise groups (from SUBSET-026-3)

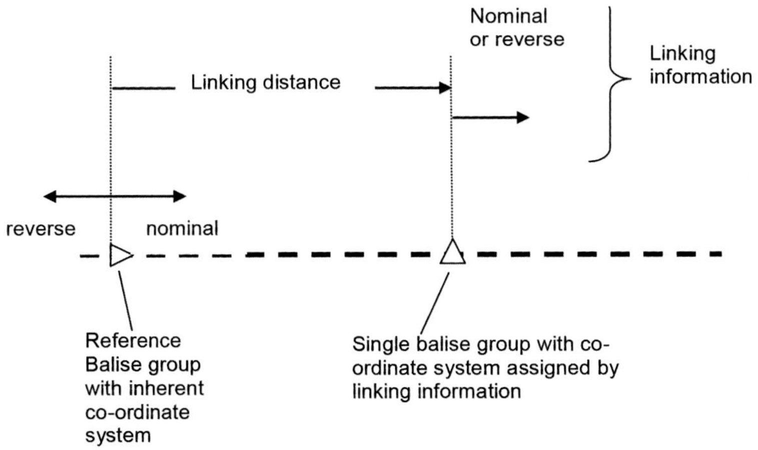

Figure 17-3: Assignment of co-ordinate system by linking (from SUBSET-026-3)

When preparing track descriptions, the distance to the location at which the next change of information (e.g. static speed profile or gradient profile change) takes effect is the incremental nominal or reverse distance from the location reference at the specified balise group. The distance values can be given in steps of 0.1, 1.0 or 10 m. The parameter Q_SCALE is used to define the resolution. The distance value given from the trackside is the nominal value with the precision defined by Q_SCALE.

The linking of balise groups is applied in ETCS to help ensure the integrity of the route data received by the onboard system. The identity, order, spacing and orientation of balise groups are specified in the track system. The linking information from a previous balise or from the RBC specifies the location accuracy of the reference balise in each balise group. If no linking information is available, a National Value is applied, or a Default Value if no National Value is applicable. Linking is also used to specify the on-board reaction (Trip, Service Brake or No Reaction) in response to balise groups that are faulty or not found. Level 2 can usually tolerate missing/faulty balise groups, so the reaction is set to No Reaction which allows the train to continue and in the background the error is logged.

It is not necessary for all balise groups to be linked. For instance, a temporary speed restriction can be applied by temporary installation of a balise group on the track without the need to amend the linking information of the other balises in the area.

17.2 Odometry

To maintain an up-to-date record of the current location of the train, the ETCS on-board system needs to know the distance and direction of travel from the last location reference balise. The ETCS standards do not mandate a particular technology for this, but in almost every case the primary system is an odometer which determines the distance moved in the direction of travel. It does this by directly recording wheel revolutions using an axle mounted sensor (usually an optical or Hall effect device). This also provides a measurement of the current speed of the train.

One significant problem with odometry is that accurate measurement of speed and distance relies on a continuous rolling contact between the wheel and the rail. In low adhesion conditions, wheelspin during acceleration or climbing a steep gradient gives an over-estimate, and wheelslide when braking gives an under-estimate, of speed and distance. One solution is to make the measurement on an axle which is unbraked and unpowered, but in most cases the consequential reduction in vehicle performance is not acceptable. For this reason, most ETCS applications incorporate a supplementary movement measurement device that is independent of wheel-rail contact, such as an accelerometer, video camera or doppler radar. The information from multiple wheel sensors and the supplementary system is then combined ('data fusion') to determine the best estimate of speed and location. To maximise availability the system may incorporate a voting algorithm to allow the detection and isolation of a faulty sensor.

Figure 17-4: Balise antenna, doppler radar and axle mounted odometer

17.3 Measurement errors and confidence interval

The distance and speed measurements of the ETCS on-board system are subject to measurement tolerances. This is due to both the inherent error of the position reference at a balise group and the accruing odometry error that accumulates between position references. There are several factors that can contribute to these errors:

Inherent error at position reference	Accruing odometry error
Installation of balise on track	Wheel diameter initial measurement
Measurement of balise signal strength	Wheel diameter wear
Clock accuracy determining time at balise	Slip between wheel and rail

Table 17-1: Position reference and odometry error factors

Each factor has its own characteristics that could contribute to location errors, some of which may also vary over time and with environmental factors. An important element of the safety case for the ETCS on-board is to demonstrate that all these factors have been taken into account in an algorithm which calculates the confidence interval for the current train location.

The confidence interval will be reset to the inherent error value of that group each time a balise group is passed, and will then progressively increase due to the accruing odometry error until the train reaches the next balise group.

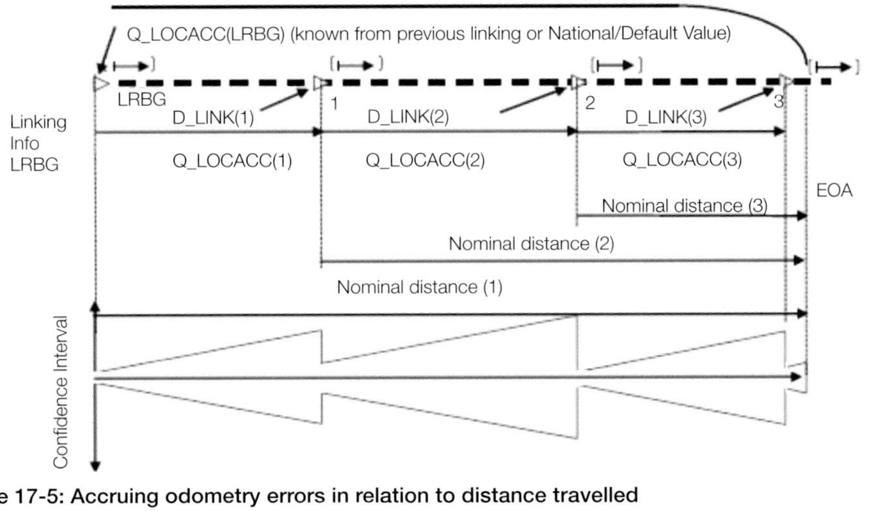

Figure 17-5: Accruing odometry errors in relation to distance travelled

The maximum acceptable value for the confidence interval depends on operational circumstances and the track layout. It becomes critical where a train needs to stop accurately at the end of a movement authority, for example at a station platform. Installation of additional balise groups may be required to reset the confidence interval on approach to these locations.

The ETCS on-board system also performs periodic checks that the odometry error does not exceed a specified percentage of the distance travelled. If the error exceeds 5% the driver is informed and if it exceeds 30% while running in either level 1 or 2 the ETCS will enter System Failure (SF) mode.

The confidence interval and train length are added or subtracted from the Estimated Location, to calculate the following values:

Maximum Safe Front End is Estimated Location plus the upper value in the confidence interval, i.e. the most forward position at which the front end of the train could already be.

Minimum Safe Front End is the Estimated Location minus the upper value in the confidence interval, i.e. the rearmost location at which the front end of the train could be.

Maximum Safe Rear End is Maximum Safe Front End minus train length, i.e. the most forward position at which the rear end of the train could already be.

Minimum Safe Rear End is Minimum Safe Front End minus train length, i.e. the rearmost location at which the rear end of the train could be.

Each ETCS on-board function then uses the appropriate location value to ensure the location accuracy is taken account of in a safe manner. For example, Maximum Safe Front End is used when supervising the end of a movement authority, but Minimum Safe Rear End is used to determine when the rear of a train has passed the end of a temporary speed restriction due to poor quality track.

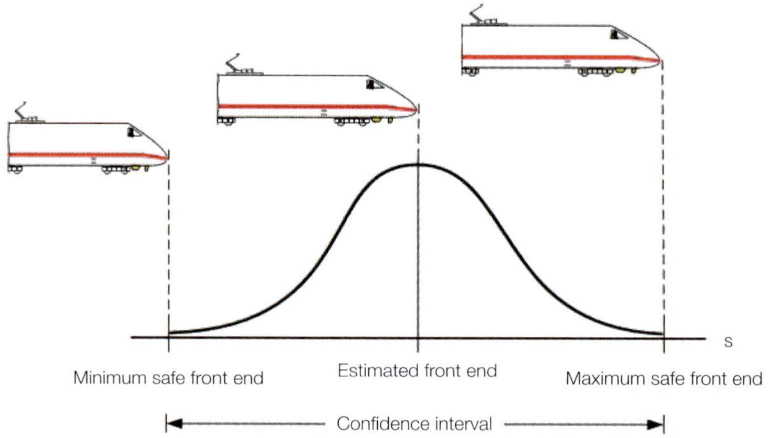

Figure 17-6: Front end in relation to the confidence interval

17.4 Repositioning

In ETCS level 1 areas, movement authority information is often derived from the feeds to the lineside signals. On approaching a junction, the signal aspect displayed may not always uniquely indicate which route is currently set, which is problematical for ETCS. For example, in the example in figure 17-7, signal A may display the same aspect for the routes to tracks 1, 2, 3.

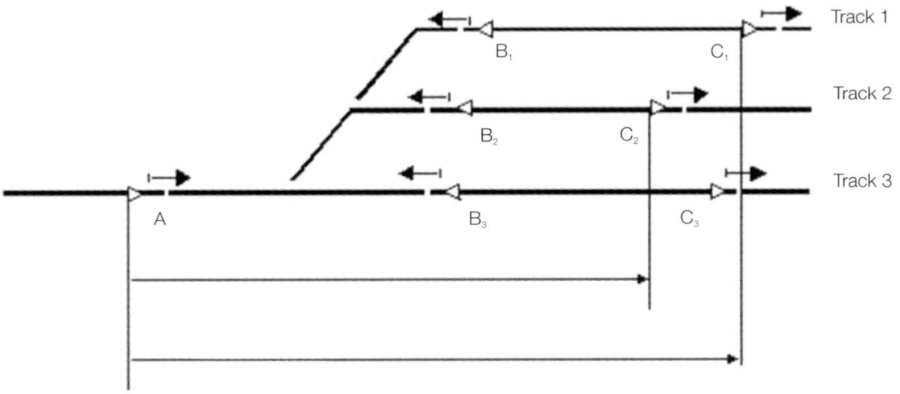

Figure 17-7: Principle of repositioning

This means that the movement authority provided by the balise group at signal A has to be a generic one that allows a train to operate safely on any of the three routes:

- The end of authority is the shortest distance to any of the three possible destinations C1, C2, C3.
- The track conditions are the most restrictive values for all the routes (could be a combination from the routes).
- The linking distance is to the furthest balise group containing repositioning information, with a special value of NID_BG which equates to 'unknown'.

The movement authority is then updated when the train passes over a subsequent balise group that uniquely identifies which track the train has been routed over. In this example these are balise groups B1, B2, B3 which are provided primarily for the signals protecting the junction but can also be used for the routing information for trains moving away from the junction. The information provided is:

- A repositioning data packet containing the distance to the end of the current section (e.g. from B1 to C1) which is used to update the end of authority location for the existing movement authority held in the ETCS on-board.
- The track description specific to this track.
- Linking to the next balise group on this track.

This can all be static information in a fixed balise as it only provides an update to an existing movement authority and will only be recognised as such by an on-board system that received the special "unknown" value of NID_BG at the previous balise group.

17.5 Cold movement detection

The train location and some other information is stored in the ETCS on-board on entering No Power (NP) mode, and it can be recovered and used in the next start of mission procedure. However, this information will only be valid if the train has not moved while the ETCS on-board is in NP mode.

This assurance is provided by a high integrity cold movement detector which can detect movement of the vehicle on which it is installed independently of the status of the ETCS on-board system. It will do this for a minimum of 72 hours after entry into NP mode. When the ETCS on-board is powered up it checks the status reported by the cold movement detector to determine whether the stored information remains valid. To allow small movements (e.g. for coupling vehicles) in NP mode, a movement of up to two metres is permissible before the stored data becomes invalid.

References

SUBSET-026-3: System Requirements Specification, Chapter 3 Principles

Chapter 18: EuroRadio Track to Train Data Interface

18.1 Introduction

For ETCS level 2 a radio data interface is required between each train's ETCS on-board and the Radio Block Centre (RBC) for the area in which the train is operating, via Railway Mobile Radio (RMR). RMR is the generic term introduced in the 2023 TSI CCS for both GSM-R, which is the current standard mandated in the European Interoperability Directives for operational voice communications to trains, as well as for ETCS data, and also FRMCS (Future Rail Mobile Communication System) which is under development to replace it.

GSM-R is the railway specific version of the '2nd generation' public mobile radio system GSM. In public networks GSM has largely been superseded by the later 3G, 4G and 5G technologies but GSM-R will continue to be supported for railway applications until at least 2030. FRMCS will be based on 5G technology.

Telecommunications is a complex and specialist subject. In this textbook we are focusing on how the fixed and mobile networks are used for the purpose of ETCS. A more detailed explanation of the engineering of GSM-R systems for other railway applications may be found in the IRSE Textbook on Railway Telecommunications.

GSM is a digital communication system, originally specified to operate only in a circuit switched (CS) mode of operation, which is analogous to a traditional point to point telephone call. To exchange data or make a voice call in CS mode a circuit is established through the mobile communications network, and a channel of the radio system is then allocated until the call is terminated. The disadvantage of this approach is that the channel remains allocated, even if there are gaps in speech or data transmission during the call. This is wasteful of capacity, and for this reason, the GSM standard was later updated to provide a packet switched (PS) mode of operation, General Packet Radio Service (GPRS). This allows packets of data to be routed through the system using the internet protocol (IP). GPRS allows the available radio channels to be used much more efficiently, as one channel can convey packets of data for multiple applications.

ETCS was originally developed to use the CS mode of GSM-R, and an option to use PS mode where the radio network supports GPRS was added in the Baseline 3 standards. An ETCS on-board system that is compliant with these standards will initially attempt to communicate using PS mode and revert to CS mode if this is not supported by the network or the RBC.

Where ETCS level 2 has been applied outside Europe, the radio frequencies required for GSM-R may not be available, and alternative radio systems such as Tetra or LTE/4G have been used instead. These variants are not standardised and therefore are outside scope for this textbook.

The use of any type of radio to communicate safety-critical information requires robust defences to protect the data from corruption and other threats. Within ETCS this is achieved using the EuroRadio protocol stack which is described in section 18.6.

18.2 GSM-R network design

Radio spectrum is a finite and valuable resource and the available frequencies are allocated by the national telecommunications regulators in each country. Railways must justify their requirement for radio channels and show that efficient use will be made of the frequencies assigned to them. It is also in the railways' interests to design their systems to make efficient use of the spectra allocated to avoid unnecessarily high costs for radio licences.

Railways need sufficient radio capacity to support the anticipated safety applications, including ETCS. Following lengthy negotiations over spectrum requirements for GSM-R, the regulators in Europe assigned a spectrum in the 900 MHz band adjacent to the public GSM band (figure 18-1). A band of 4 MHz in each of the uplink (mobile to fixed network) and downlink bands assured the long-term availability of a common set of frequencies across the EU, for all railways to use for the range of on-board applications required to meet the interoperability directives. Within each 4 MHz band (uplink and downlink) there are 19 x 200 kHz usable frequency bands, each one capable of carrying seven voice/data channels.

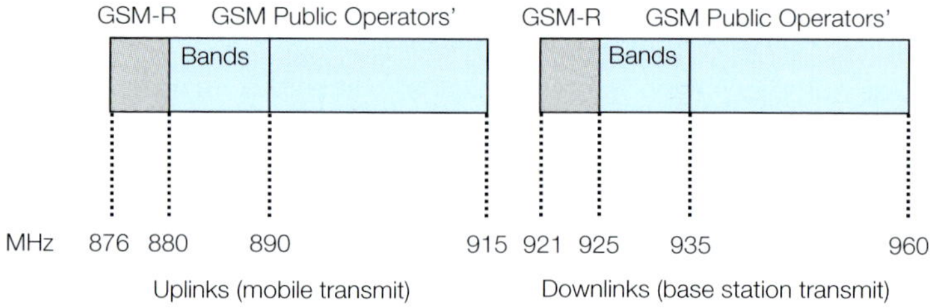

Figure 18-1: Frequency assignments in the GSM bands

GSM-R is a cellular mobile communications system whose radio transmitters are placed a few kilometres apart to give blanket coverage of the railway. Neighbouring transmitters use different 200 kHz frequency bands to avoid interference, and their areas of geographical coverage (the 'cells') are designed to overlap so that a connection can be maintained between the train and the RBC as the train crosses the boundary between one transmitter and the next. This overlap distance will depend on the maximum speed allowed (up to the defined system maximum of 500 km/h) and on the time required by the system to recognise the changes in field strengths in the overlap area (which triggers the handover to the next transmitter) and to complete the handover process. This requires approximately six seconds, of which the actual handover process takes approximately one second.

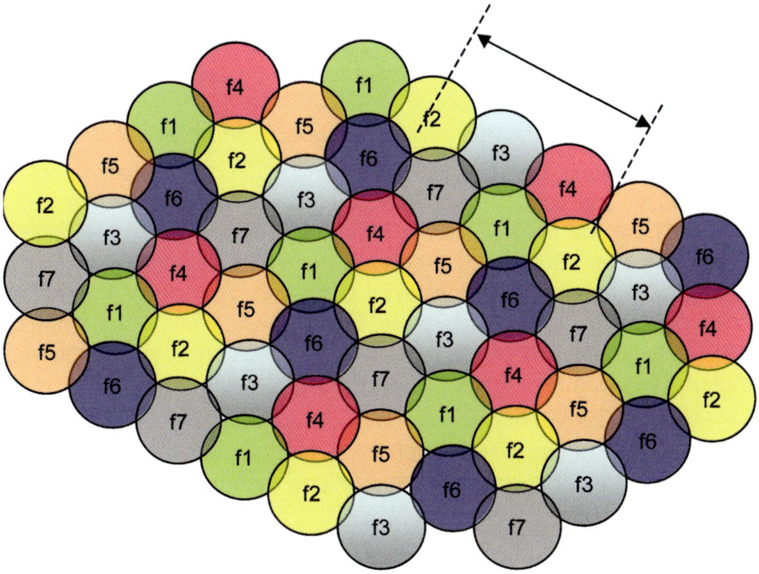

Figure 18-2: Frequency allocation in a cellular network

Figure 18-2 shows how seven frequencies can be allocated over a geographical area with a spacing of at least two cells between cells of the same frequency to limit co-channel interference. The spectrum allocated to the railways for GSM-R only allows for 19 x 200 kHz bands, so typically no more than two bands can be allocated to each cell. Where CS mode is used, each band can carry seven voice/data channels simultaneously. There are theoretically eight channels available, but one is used as a control channel. Depending on the level of traffic, it may be possible to use one control channel for two frequency bands in a cell, thus allowing a maximum of 15 voice/data channels over the two frequency bands.

In practice, planning a railway GSM-R network is rather different from the classic cellular layout. Most of the radio coverage area is very long and narrow, except in large conurbations and cities. Furthermore, the radio must work in tunnels, deep cuttings and urban areas with buildings, bridges and catenary that may obstruct the radio signals. By increasing the height of transmitters and focusing the transmission along the track with narrow beam antennas, inter-site separation distances in the range 10-15 km can be obtained, with improved separation in urban areas and significant cost reduction. Planners generally use a software model coupled with detailed digital maps with 20-50 metres resolution to deliver the required field strength to a given level of probability over the area to be covered.

To manage the limited capacity of GSM-R, one of its railway specific features is that it allows a priority to be assigned to each call, depending on the relative operational importance (0 = highest to 4 = lowest). The most important is a railway emergency call, which is the only call type that has a priority of 0. The system guarantees that a railway emergency call is successful by terminating and preventing calls of a lower priority if necessary (all channels busy). An ETCS data call has a priority of the next rank of 1. This is a high priority operational call, but it could be pre-empted by a railway emergency.

Even with call prioritisation, the capacity limit could become a problem when ETCS is used in a busy station area, because in CS mode each train needs a dedicated data channel. The solution is to use the PS mode of communication which allows data channels to be shared by many trains.

18.3 Quality of Service

An important element of the research and standardisation process for ETCS has been to establish Quality of Service (QoS) standards for GSM-R networks that can be guaranteed to deliver reliable operation and are both practically achievable and cost effective.

ETCS is able to work reliably using circuit switched data in an unreliable data transmission environment because the data messages are short. Position data and message acknowledgements from the train are about 14 bytes, and movement authority data to the train varies between 40 and 500 bytes. However, the transfer of data is time critical, so it is important that the data transmitted is received both with a predictable error rate and with delays that are low enough to be acceptable. By contrast, these effects have less impact on a voice call because there is ample redundancy in speech. The listener will ask for the speaker to repeat what was said if any part of the message is not heard or may have been misunderstood due to corruption of the signal.

The safety-critical nature of the ETCS application places unusually high demands on the quality of service required of the data communications system. For CS mode this translates into a user expectation of:

- A high call success rate
- A low data block error rate
- A low rate of dropped calls (the calls might need to be sustained for hours rather than minutes, depending on train journey time).

GSM technology was not originally designed to meet the demands placed upon it by ETCS applications. Reliable operation of ETCS level 2 is dependent on the GSM-R network being engineered to deliver a high QoS and this may require smaller cell areas or additional radio antennas compared with a network that is adequate for voice calls. The ETCS requirements for GSM-R in CS mode are defined in SUBSET-093.

When PS mode is used, the QoS requirements are different but equally demanding. ETCS may be sharing the network with other applications that are transferring much larger volumes of data, and the ETCS packets need to be given sufficient priority to ensure timely data exchange to and from trains. The QoS requirements for PS mode are specified in UIC (International Union of Railways) document O-8664-1.0.0.

18.4 Interface on the train

The interface between the ETCS on-board and the GSM-R radio system is provided by a GSM-R data radio (referred to as the 'mobile equipment') connected to an antenna on the roof of the train. There are usually two radios so that when a train crosses a boundary between two RBC areas, or between GSM-R network areas, a new connection can be established while the old connection continues in use. The ETCS data radios are independent of the GSM-R cab radio provided for voice communications between the driver and the control centre.

The core of the data radio is just the same as any mobile phone, with a SIM card that defines its identity to the GSM-R network. For international trains, it will be capable of 'roaming' between different infrastructure managers' GSM-R networks.

The interface between the ETCS on-board and the mobile equipment is referred to as IGSM and is specified in the Radio Transmission FFFIS for EuroRadio (UIC A11T 6001 13.0.0).

At the hardware level this uses the ITU V.24 interface (RS-232) with asynchronous serial communications at 9600 baud for CS operation and 19200 baud or greater for PS operation.

18.5 Interface at the control centre

The interface between the RBC at the control centre and the GSM-R radio system is provided by the infrastructure manager's fixed telecommunications system. The interface for circuit switched (CS) operation is referred to as IFIX_CS and is specified to use the ISDN (Integrated Services Digital Network) primary rate user network interface at 2048 kbps, which allows an RBC to manage 30 simultaneous connections.

The interface for packet switched (PS) operation is referred to as IFIX_PS, which uses Ethernet hardware and IPv4 (Internet Protocol version 4). Again, the detailed requirements are specified in the Radio Transmission FFFIS for EuroRadio. An RBC that is capable of both CS and PS operation will have both types of interface, and each of these will usually be duplicated for availability.

18.6 EuroRadio protocol stack

Modern communications are based on the ISO Open Systems Interconnection (OSI) protocol stack model, a conceptual framework that divides network communications functions into several layers (the 'stack'), enabling 'applications' to communicate without knowledge of the hardware and software that they are using. At the sending end of a data link, information is passed down the stack from the application layer to the physical layer, with each layer embedding the information received from the layer above in the data that it passes to the layer below. At the receiving end, each layer strips out the data specific to its own layer and passes the embedded information upward. This allows the software in each layer of the stack to communicate independently of the layers above and below.

The ETCS implementation of the OSI model has six layers (figure 18-3). The top layer is the ETCS application, i.e. the ETCS on-board and RBC functionality, and the bottom 'physical' layer is the GSM-R radio system (but note that within the GSM-R system is another protocol stack). 'EuroRadio' is the name given to the four intermediate layers. Its role is to protect safety-related data from corruption and other threats when it is transmitted over the radio system. EuroRadio is usually implemented as a software module within the ETCS on-board and RBC, and it is fully specified in the EuroRadio FIS SUBSET-037.

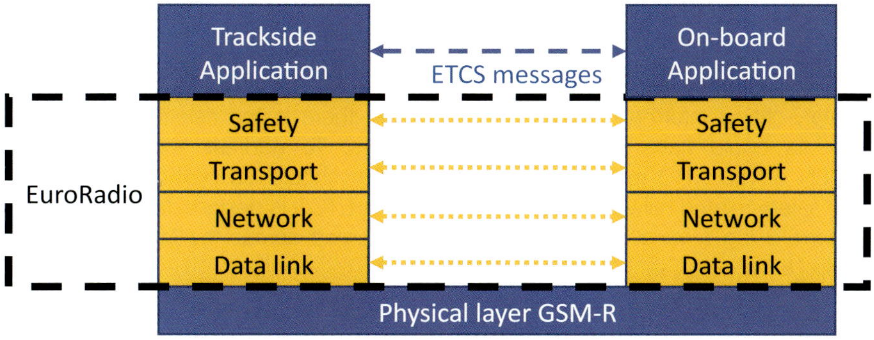

Figure 18-3: EuroRadio protocol stack

The top layer within EuroRadio is the safety layer. This is responsible for authentication (confirming the identity of the two applications involved in the communication) and it also acts as a final error-checker. It is the guarantee of most of the communications safety and so it is generally implemented at SIL4, the highest safety integrity level for railway applications. Unlike layers below, it is not a standard protocol – it has been developed specifically for ETCS purposes.

The three lower layers within EuroRadio (transport, network and data link) all make use of standard communications protocols which are necessary for circuit switched (CS) and packet switched (PS) data. This includes the messages needed in each layer to establish and release connections through the lower layers, and segmenting and reassembling longer messages (see figure 18-4). This latter feature is needed as the application information can be up to 1023 8-bit bytes in length, but the HDLC (high-level data link control) frames sent to the physical layer are limited to 41 bytes. The software in these layers also includes the co-ordination functions to determine the choice of CS and PS modes of operation. These layers also carry out extensive error checking that the data has been transmitted and received correctly, but they are nevertheless treated as part of the 'untrusted' communications system.

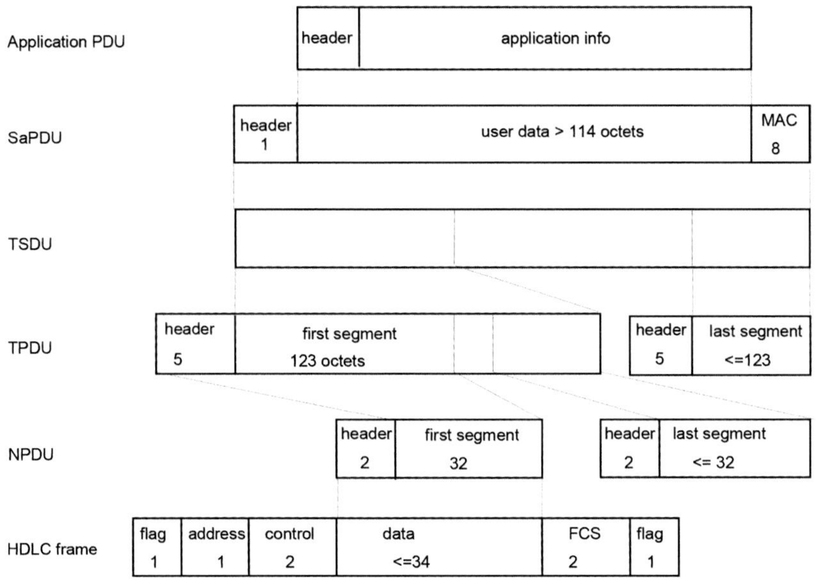

Figure 18-4: Segmenting/reassembling of messages in EuroRadio (from SUBSET-037-1)

Communications of any sort, and especially by radio, are vulnerable to a number of threats. Where safety-critical information is being exchanged there is the risk that any error in the received data could have hazardous consequences. To address this issue, CENELEC standard EN 50159:2010 recommends defences that should be provided against credible threats for safety-related railway signalling applications. ETCS is designed to be compliant with this standard, using a variety of defences that are spread across the layers of the protocol stack (table 18-1).

Protocol Layer	Defences incorporated	Threats addressed
Application layer	Time stamp Timeout Source/destination ID Feedback messages	Repetition Insertion Re-sequence Delay
Safety layer	Source/destination ID Safety code Cryptographic techniques	Corruption Masquerade
Transport layer	Source/destination ID Sequence number (where long message split)	Repetition Deletion Insertion Re-sequence
Network layer	Source/destination ID Sequence number	Repetition Deletion Insertion Re-sequence
Data link layer	Sequence number Safety code Feedback message Timeout	Repetition Deletion Insertion Re-sequence Corruption Delay

Table 18-1: Examples of EN 50159 defences against threats

18.7 Message authentication and key management

The most challenging threat to safety communication is that of a 'masquerade', where an apparently genuine message has been fabricated or modified deliberately and maliciously. EuroRadio provides a defence against this in the safety layer by appending a message authentication code (MAC) to the end of each message to provide error detection and conformation of authenticity. The MAC is generated by a particular form of encryption that produces a code from all the bits of the message and a secret 'session key' (KSMAC). The message itself is not encrypted because in railway signalling applications the information does not need to kept secret, it just needs to be proved genuine, which is possible provided both sender and receiver have access to the same KSMAC.

The basic encryption method is the Data Encryption Standard (DES) originally developed in the 1970s. It uses a 64-bit key, of which 56 bits are the actual random key, and 8 are parity bits. An effective key of 56 bits is rather small considering the processing power now available. This was recognised when EuroRadio was first specified, and the actual algorithm used is called 'single DES with modified MAC algorithm 3'. This is a mechanism for giving the strength of triple-DES (which is used extensively in the electronic payments industry) without incurring the extra time involved in triple-encrypting all blocks of data. This means that the KSMAC is actually 192 bits, containing three independent DES keys, and gives an effective key length of a 112-bit key system.

Whenever a connection is first established between a train and an RBC, a new KSMAC is generated using a stored 'authentication key' (KMAC) and two random numbers, one generated on the train and one at the RBC. This is achieved by a sequence of four authentication messages between the safety layers as part of the setting up a connection from the train to the RBC (figure 18-5). If a hacker was able to monitor the communications, they would only see the random numbers and without the KMAC would not be able to deduce the

KSMAC to be used in the connection. A fresh KSMAC is generated for each new connection which gives a short time window in which to attempt to break the encryption and discover the KSMAC in use. Timestamps on messages that increment all the time the connection is established also make it impossible for a hacker to record a message and replay it later.

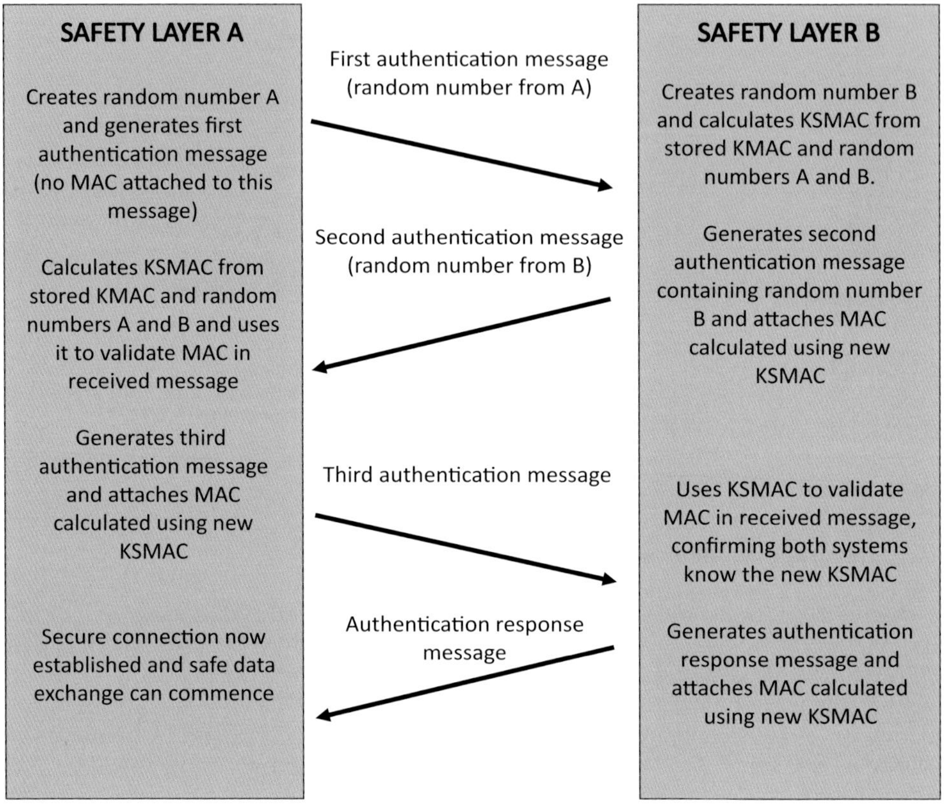

Figure 18-5: Time sequence during connection establishment

This approach does however rely on the train and the RBC having the same stored KMAC, which leads to a requirement for key management. Each ETCS entity that communicates via EuroRadio must store a list of keys. Each key comprises a unique identifier issued by a key management centre (KMC), a KMAC value, a validity period for the key, and a list of peer ETCS entities that the key can be used to communicate with (for an RBC a list of trains, for an on-board system a list of RBCs). This provides flexibility in the way that keys can be used – the same key could be used for every train and RBC in a large area for an indefinite period, or a unique key defined for every on-board/RBC pair, updated every few days. There is clearly a balance between the level of security that is achieved and the overhead effort of updating lists of keys in every entity.

The creation, deletion and updating of keys is undertaken by the KMCs which perform an essential role when new ETCS entities (e.g. trains) are placed into service or if the area of operation of a train is amended, and in managing updates to keys that are required to maintain the security of the system against cyber-threats. Every ETCS entity that needs to communicate via EuroRadio has to be associated with a specific 'home' KMC, and can only accept keys from this source. When a train operates outside its home KMC area, the necessary keys and lists of peer ETCS entities must first be exchanged between the respective KMCs so that they are available to the train and all the RBCs it will encounter.

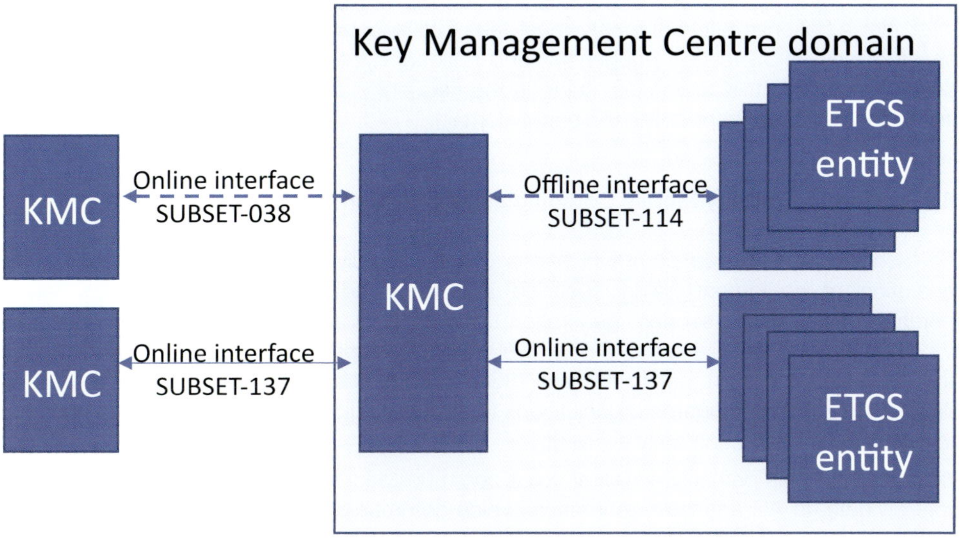

Figure 18-6: Key management interfaces and specifications

The ETCS standards define two approaches to key management. Where 'online' key management is used, the ETCS entity has a data communication link to the KMC, while for 'offline' key management, a physical device such as a memory stick is used to transfer the keys. In both cases there is an additional key KTRANS which has two parts. The first is used to authenticate the key management message, and the second to encrypt the key being distributed. The same applies to messages between KMCs, where the additional key is known as K-KMC.

Offline key management is specified as a FIS (specifically, SUBSET-114 for KMC to ETCS entity; and SUBSET-038 for KMC-KMC) which defines the data to be exchanged and its encoding, but which does not constrain the physical media used to transfer the information.

Online key management is a fully interoperable interface that is defined in a FFFIS (SUBSET-137). This is a TCP/IP (transmission control protocol/internet protocol) interface with an additional security layer known as TLS (transport layer security) using public key infrastructure (PKI) technology to allow the use of any fixed or mobile communication network.

Each KMC has an ETCS ID which uniquely identifies it. The first part of the ID is a NID_C value that defines a country or region, and the second part is a number allocated by the national authority responsible for that geographical area. Within these constraints there is no limit to the number of KMCs and the size of their 'domains', but the most common approach will be to have a single KMC operated by the Infrastructure Manager for a NID_C area, which manages keys for all the RBCs in the area and all the trains that are based within the area. However, other solutions are possible – for instance a KMC could be managed by an international train operator, a rolling stock owning company or an equipment supplier, communicating with the KMCs of the infrastructure managers over which the trains will operate.

Any organisation operating a KMC will need to have robust operating procedures, the necessary computer hardware and software, and strong physical and cyber-security measures. The security and availability of the key management service will be a critical factor that a national safety authority must take into account when authorising the entry into service of any ETCS-related equipment, and national cyber-security organisations may also have an involvement. Key policies, such as the frequency of key updates, will be confidential information which are kept out of the public domain.

18.8 Future Railway Mobile Communications System

The technology of mobile communications evolves rapidly, and each decade sees a new generation of equipment being specified and deployed by the public mobile network operators. GSM is a second generation (2G) technology which has been succeeded by 3G, 4G, and 5G systems. Since 1999 the development has been co-ordinated by the 3rd Generation Partnership Project (3GPP) which has brought together the world's telecommunications standards organisations to achieve a global standard. Over this period the standards have been constantly evolving.

Whereas GSM was primarily aimed at voice communications, the later standards focus much more on data. In 4G and later generations voice communication is simply an application that converts speech to data packets that are passed over the network alongside other data. This is known as VoLTE (Voice over Long-Term Evolution). Together with the opening up of additional higher frequency bands in the radio spectrum, this enables much higher data communication rates and lower latency.

Until recently most public telecommunications operators have operated a mixture of generations of equipment, with older systems remaining in rural areas while the higher capacity modern systems have been installed in urban areas where there is a high data demand from users. This era is coming to an end as older systems are switched-off, leaving the railways' GSM-R systems as the last users of 2G technology. Equipment suppliers have given an undertaking to support GSM-R until 2030, but the railways have had to develop a strategy for migrating to a next generation technology thereafter.

The replacement for GSM-R is known as the Future Railway Mobile Communication System (FRMCS) and will be based on 5G technology. To this end railways and suppliers have been participating in the 3GPP developments to ensure that requirements for railway applications are taken into account in the evolving standards. Fortunately, the requirements that were 'add-ons' to the basic GSM specifications for GSM-R are now an integral part of the 3GPP specifications, because they are relevant to a wider range of users. For instance, 'mission critical' voice and data communications are a requirement for public security and emergency services, and low latency data communications may be required for autonomous road vehicles.

The railway specific FRMCS standard specifies a standard approach to using the generic 3GPP standards in a manner that will ensure interoperability. The standards framework for this is defined in the 2023 TSI CCS which has introduced the term Railway Mobile Radio (RMR) as the generic term for both GSM-R and FRMCS. However, the detailed content of the FRMCS standard will only be finalised in a later TSI CCS after a number of research and demonstration projects have been completed.

An important consideration for infrastructure managers and railway undertakings will be the strategy for migration from GSM-R to FRMCS, as some level of dual fitting of infrastructure and/or trains will be required for a transitional period. The 2023 TSI CCS establishes some basic rules for this. Infrastructure managers must give a minimum of 5 years notice to railway undertakings before they can switch off GSM-R and require all trains to be fitted with FRMCS on any given route.

Availability of frequency spectrum will be particularly critical during the migration period when GSM-R and FRMCS may be operating simultaneously in the same area. To facilitate this the EU railways have been able to secure 1.4 MHz extensions to each of the existing 4 MHz uplink and downlink frequency bands allocated for GSM-R (EU Directive 2021/1730). This will allow FRMCS to be initially deployed alongside GSM-R on existing base stations using the extended frequency bands and then take over the full frequency band at a later date when GSM-R can be switched off. An additional frequency band at 1900-1910 MHz has also been allocated for FRMCS, but full utilisation of this may require some additional base stations to achieve reliable operation at the higher frequency. It may also be possible for FRMCS to use capacity on public networks using facilities available in 5G to guarantee security and quality of service for critical applications.

The greatly increased capacity of FRMCS compared to GSM-R is seen as an enabler for many new applications that require data exchange with trains. The on-board architecture is based on the concept of an FRMCS Gateway which sits between the on-board radio (or multiple radios for different frequency bands) and applications such as ETCS, operational voice communications and additional services such as condition monitoring, customer information and video.

18.9 EuroRadio provision for FRMCS

From an ETCS point of view, FRMCS provides an alternative to using GPRS to provide packet switched (PS) communications between the on-board and trackside subsystems. The main complication will be migration. To operate on an FRMCS fitted route a train already fitted with GSM-R will require an additional FRMCS radio and gateway to be installed, and the ETCS on-board will need to be able to select the appropriate mode of operation for the line on which it is running. This is basically an extension of the existing functionality that selects between CS and PS modes of operation on GSM-R. To cover all the eventualities, at Baseline 4 the EuroRadio standard SUBSET-037 has been structured in three parts:

- SUBSET-037-1 EuroRadio FIS GSM-R CS/PS Communication Functional Module and Coordinating Function FRMCS/GSM-R
- SUBSET-037-2 EuroRadio FIS Safety Layer
- SUBSET-037-3 EuroRadio FIS FRMCS Communication Functional Module

The cyber-security threat is constantly evolving, and it is recognised that the level of encryption provided by EuroRadio is becoming inadequate in the face of modern computing power. The use of IP protocols and sharing of the FRMCS communication channel with other applications also increases the risk of a cyber-attack. For this reason, the interface to FRMCS will introduce a new security layer, which will be used in addition to the existing EuroRadio safety layer for the on-board ETCS to RBC interface and also for new applications such as ATO. The new security layer is specified in SUBSET-146 ERTMS End-to-End Security. This is based on the principles previously adopted for online key management, transport layer security (TLS) and public key infrastructure (PKI) technology which allows more modern encryption coding that can be readily updated as new threats emerge. The relationship between this additional security layer and the three EuroRadio SUBSETs is shown in figure 18-7.

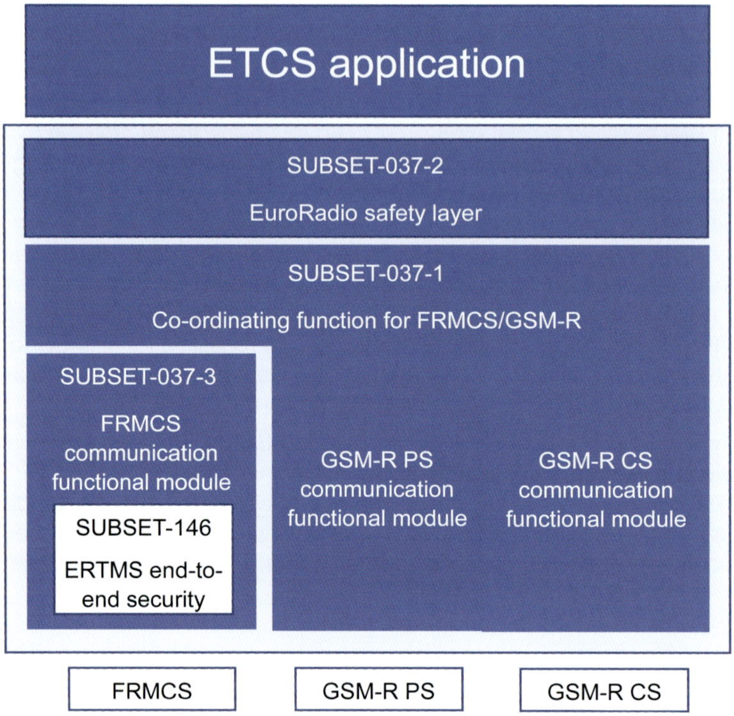

Figure 18-7: Content and relationships between EuroRadio specifications for Baseline 4

References

SUBSET-037-1: EuroRadio FIS GSM-R CS/PS Communication Functional Module and Coordinating Function FRMCS/GSM-R

SUBSET-037-2: EuroRadio FIS Safety Layer

SUBSET-037-3: EuroRadio FIS FRMCS Communication Functional Module

SUBSET-038: Offline key management FIS

SUBSET-093: GSM-R Bearer Service Requirements

SUBSET-114: KMC-ETCS Entity Off-line KM FIS

SUBSET-137: On-line Key Management FFFIS

SUBSET-146: ERTMS End-to-End Security

UIC O-8664-1: ETCS in PS-mode GPRS/EGPRS Guideline

UIC A11T 6001: Radio Transmission FFFIS for EuroRadio

CENELEC EN 50159:2010: Railway applications - Communication, signalling and processing systems - Safety-related communication in transmission systems

IRSE textbook: Railway Telecommunications (2004), ISBN: 9780902390171

Chapter 19: ETCS Message Language and Structure

19.1 Introduction

This chapter explains the definition of the messages by which data is exchanged across system interfaces and the message structure, coding and checking process. It also describes how data is arranged before transmission and the pre-transmission checks that are performed.

The ETCS system uses its own language to exchange data between the trackside and the on-board assemblies. Specifically, it is used for transmitting information over the EuroRadio and Eurobalise airgaps, and between the ETCS on-board, STM and JRU. The language is based on variables, packets, messages, and telegrams, and all the definitions of these elements are independent of the transmission media used. The language is designed to minimise the volume of data exchanged by encoding the information as a binary bit string.

An ETCS message comprises a sequence of variables and packets that are sent between the on-board and trackside applications. The message is embedded within a safety protocol which is specific to the application medium used to transmit the data (EuroRadio, Eurobalise) – see chapters 16 and 18 for more information about the protocols.

19.2 Variables

Variables are used to encode single data values. Variables cannot be split into smaller units; the whole variable has a single meaning and is encoded as a fixed length binary number. The size of the variables varies from 1 to 64 bits and when assembled into a message the most significant bit of each variable is transmitted first. The value of a variable can either represent a numerical item such as time or distance or can have a special value related to the basic meaning of the variable.

The names of all variables are unique. All variable names start with a prefix which defines the type of variable.

A_	Acceleration		NC_	Class number
D_	Distance		NID_	Identity number
G_	Gradient		Q_	Qualifier
L_	Length		T_	Time/date
M_	Miscellaneous		V_	Speed
N_	Number		X_	Text

Table 19-1: Variable prefixes

The full definition of each variable can be found in chapter 7.5 of the System Requirements Specification (SRS) (SUBSET-026). Some examples are given in tables 19-2 to 19-5.

SRS 7.5.1.3	D_CYCLOC		
Name	Distance between two location reports from the train		
Description	The train has to report its location every D_CYCLOC meters.		
Length of variable	Minimum Value	Maximum Value	Resolution/formula
15 bits	0 cm	327.660 km	10 cm, 1 m or 10 m depends on Q_SCALE value in the same packet
Special/Reserved Values	1111 ... 1111	The train has not to report cyclically its location.	

Table 19-2: Example of a Distance variable definition

SRS 7.5.1.49	L_PACKET		
Name	Packet length		
Description	L_PACKET indicates the length of the packet in bits, including all bits of the packet header.		
Length of variable	Minimum Value	Maximum Value	Resolution/formula
13 bits	0	8191	1 bit

Table 19-3: Example of a Length variable definition

SRS 7.5.1.69	M_LOC		
Name	Special location/moment where the train has to report its position.		
Description			
Length of variable	Minimum Value	Maximum Value	Resolution/formula
3 bits			
Special/Reserved Values	000	Now (The position report is sent upon receipt of the order.)	
	001	Every LRBG compliant balise group.	
	010	Do not send position report on passage of LRBG compliant balise group.	
	011-111	Spare	

Table 19-4: Example of a Miscellaneous variable definition

SRS 7.5.1.129	**Q_SCALE**		
Name	Qualifier for the distance scale		
Description	Qualifier to indicate the same scale used for describing all distances/lengths inside the packet that contains Q_SCALE.		
Length of variable	**Minimum Value**	**Maximum Value**	**Resolution/formula**
2 bits			
Special/Reserved Values	0	10 cm scale	
	1	1 m scale	
	2	10 m scale	
	3	Spare	

Table 19-5: Example of a Qualifier variable definition

For some variables, specific values are not defined in the standard and these need to be allocated by a railway administration or operator. Examples of this are NID_ENGINE which is the unique identity assigned to the ETCS on-board system of each locomotive or train, and NID_NTC which identifies a specific variety of national train control system that may be interfaced via an STM. The allocation of values for variables such as these is overseen by the European Agency for Railways (ERA), which co-ordinates the process by assigning ranges of values to each administration as described in chapter 14. A list of the ranges that have been assigned is published regularly in the document ERA_ERTMS_040001 "Assignment of values to ERTMS variables". It is then the responsibility of each administration to allocate specific values within their assigned ranges.

19.3 Packets

Packets are comprised of multiple variables grouped into a single unit, with a defined structure. The packet definition is independent of the transmission media used. Most packets are used for track to train communication, some are specific to one type of transmission medium (EuroRadio or Eurobalise) while others are used in both. A smaller number of packets are used for train to track communication (EuroRadio from ETCS on-board to RBC). One packet ('end of information') is used in both directions.

The structure for track to train packets consists of a unique packet number, the packet length, the orientation information, the distance scale (optional), and the information section which contains the variables. The structure for train to track packets is the same as the structure for track to train packages except that 'direction' is not used. In some packets, one or more variables may be optional or repeated; these variables are preceded by a special variable such as N_ITER (number of iterations) or Q_ENDTIMER (set to 1 if end timer data follows).

The full definition of each packet is contained in a tabular format in chapter 7.4 of the SRS. Two examples are given here – see tables 19-6 and 19-7.

SRS 7.4.2.15	Track to train packet 58: Position Report Parameters		
Description	Provides the parameters telling the on-board when and how often the position has to be reported.		
Transmitted by	RBC		
Content	Variable	Length	Comment
	NID_PACKET	8	Packet identifier = 58
	Q_DIR	2	Validity direction of the data
	L_PACKET	13	Number of bits in the packet
	Q_SCALE	2	Qualifier for distance scale used in packet
	T_CYCLOC	8	Time interval between position reports
	D_CYCLOC	15	Distance between position report
	M_LOC	3	Position report immediately or passing balises
	N_ITER	5	Number of iterations of the following data set
	D_LOC(k)	15	Distance to make a position report
	Q_LGTLOC(k)	1	Qualifier for position report front/train

Table 19-6: Definition of packet number 58

Track to train packet 58 (see table 19-6) allows the RBC to specify to the train when and where it should provide position reports. The variables in the message allow reporting to be specified in accordance with the following options:

- periodically in time – the time is specified in T_CYCLOC
- periodically in space – the distance is specified in D_CYCLOC, using scaling specified in Q_SCALE
- at every balise group – M_LOC = 001
- immediately – M_LOC = 000
- when the front or rear of train passes one or more specified locations – M_ITER = number of locations; D_LOC = distance of location from last referenced balise group specified in the current message (using scaling specified in Q_SCALE); Q_LGTLOC = 0 rear of train, = 1 front of train.

SRS 7.4.2.17	Track to train packet 65: Temporary Speed Restriction		
Description	Transmission of temporary speed restriction		
Transmitted by	Balise or RBC		
Content	**Variable**	**Length**	**Comment**
	NID_PACKET	8	Packet identifier = 65
	Q_DIR	2	Validity direction of the data
	L_PACKET	13	Number of bits in the packet
	Q_SCALE	2	Qualifier for distance scale used in packet
	NID_TSR	8	Identity number for the TSR
	D_TSR	15	Distance to beginning of temporary speed restriction
	L_TSR	15	Length of the temporary speed restriction
	Q_FRONT	1	Qualifier for validity end point
	V_TSR	7	Permitted speed for the temporary speed restriction

Table 19-7: Definition of packet number 65

Track to train packet 65 (see table 19-7) is used in a message to define a temporary speed restriction. It can be sent either from a balise or from an RBC. The variable NID_TSR is provided so that a TSR can be amended or revoked in a later message, and Q_FRONT defines whether the speed restriction is to be applied until the front of the train (no train length delay) or the end of the train (train length delay) has reached the specified end location of the TSR. If the message is sent from a balise, the distance to start of the speed restriction D_TSR is from the balise location. If the message is from an RBC, the distance is from the last referenced balise group specified in the same message.

19.4 Packet 44

Packet number 44 has been allocated to allow the ETCS radio and balise communication channels to pass messages other than the standard ones required for ETCS. This facility has been used for a wide range of applications such as speed supervision for tilting trains, controlling the safe opening of train doors at stations and national (Class B) train protection systems. The flexibility provided by this feature has also allowed the early implementation of new features such as Automatic Train Operation before they have been incorporated into the international standard.

When the ETCS on-board receives a packet 44 in a balise or radio message, it is simply passed on to be processed outside the interoperable functionality provided by the ETCS on-board system. This could be an additional software module within the on-board system, a STM (see chapter 21) or an external system. If there is no on-board functionality linked to a specific packet 44 it will simply be ignored.

The effective use of packet 44 requires that each application is defined and implemented with rigour and discipline. The safety and interworking of packet 44 requires attention to the following:

- ETCS message syntax
- packet 44 message syntax
- packet 44 parameter definitions to avoid interference between applications
- safety during data changes
- whole life cycle management

Within each packet 44 there is an identifier (NID_XUSER) to indicate the organising body responsible for that packet. The allocation of these values to organising bodies is managed by European Union Agency for Railways (ERA). It is then the responsibility of each organising body to specify how the remainder of the data within the packet is organised. For instance, one of the national organising bodies is the United Kingdom Rail Safety and Standards Board, and they publish a specification for the contents of packet 44 where NID_XUSER = 9 in their standard RIS-0784-CCS "The Management of Packet 44 Applications".

19.5 Messages in Eurobalise telegrams

In the case of Eurobalises, a message is the information sent by a balise group. It may be sent in one or more telegrams, where each telegram is transmitted by an individual Eurobalise. The order in which the information is transmitted respects the order of the data elements listed in the message format (from top to bottom). However, with one exception (when infill information is transmitted), the behaviour of the onboard system does not depend on the received sequence of the packets that comprise the message.

Two balise telegram sizes are defined. A short telegram consists of 341 bits, of which 210 are available for user data (the remaining bits are used for the protocol and safety encoding). A long telegram consists of 1023 bits, of which 830 are available for user data. The telegram header consists of 50 bits and the packet 'End of Telegram' of 8 bits. This leaves 152 and 772 bits respectively for the data packets. If a longer message is required at a particular location, a balise group with additional balises is required so that the message can be split across multiple telegrams, one being transmitted by each balise. The user data transmitted by an individual balise must contain complete packets – splitting a packet between two balises is forbidden.

Variable	Length	Remarks
Q_UPDOWN	1	Defines the direction of the information
M_VERSION	7	Version of the ERTMS/ETCS system
Q_MEDIA	1	Defines the type of media as balise (0)
N_PIG	3	Position in the group, defines the position of the balise in the balise group
N_TOTAL	3	Total number of balises in the balise group
M_DUP	2	Used to indicate whether the information of the balise is a duplicate of the balise before or after this one
M_MCOUNT	8	Message counter to enable detection of a change of balise group message during passage of the balise group
NID_C	10	Country or region
NID_BG	14	Identity of the balise group
Q_LINK	1	Marks the balise group as linked or unlinked
Packet 0	14	Virtual Balise Cover marker (optional)
Packet(s)	Variable	One or more packets of data making up the part of the message sent from this balise within the group
Packet 255	8	End of Information

Table 19-8: Eurobalise telegram content

The packet definitions in Chapter 7 of the ETCS SRS state which packets can be sent in a balise telegram, and Chapter 8 defines the rules for which combination of packets are permitted.

19.6 Messages via EuroRadio

Every EuroRadio message starts with a message identifier, which specifies the structure of the variables and packets that make up the message. Chapter 8 of the ETCS SRS defines the permitted messages and their usage, i.e. track to train or train to track. For some messages, it is possible to add optional packets at the end of the message.

Message formats are shown in tables 19-9 and 19-10, and message types in tables 19-11 and 19-12.

Variable	Length	Remarks
NID_MESSAGE	8	Message Identification Number
L_MESSAGE	10	Message length
T_TRAIN	32	Time Stamp from RBC
M_ACK	1	Acknowledgement required/not required from on-board
NID_LRBG	24	Identification Number of LRBG
Variables	Variable	Variables not in a packet, required in this message
Packets	Variable	Packets required in this message
Optional packets	Variable	As permitted with this message
Padding	Variable	If required to make the message length up to an integer number of octets as required by the protocol layers

Table 19-9: Standard format of EuroRadio track to train message

Variable	Length	Remarks
NID_MESSAGE	8	Message Identification Number
L_MESSAGE	10	Message length
T_TRAIN	32	Time Stamp from Train
NID_ENGINE		Identity of the train
Variables	Variable	Variables not in a packet, required in this message
Packet 0 or 1	Variable	Position report (except in messages 146,154,155,156,19)
Packets	Variable	Packets required in this message
Optional packets	Variable	As permitted with this message
Padding	Variable	If required to make the message length up to an integer number of octets as required by the protocol layers

Table 19-10: Standard format of EuroRadio train to track message

ID	Message Name
2	SR Authorisation
3	Movement Authority
4	SM Authorisation
5	SM Refused
6	Recognition of exit from TRIP mode
7	Acknowledgement of safe consist length info for SM
8	Acknowledgement of Train Data
9	Request to Shorten MA
15	Conditional Emergency Stop
16	Unconditional Emergency Stop
18	Revocation of Emergency Stop
24	General message
27	SH Refused
28	SH Authorised
32	RBC/RIU System Version
33	MA with Shifted Location Reference
34	Track Ahead Free Request
37	Infill MA
38	Acknowledgement of session establishment
39	Acknowledgement of termination of a communication session
40	Train Rejected
41	Train Accepted
43	Start of mission position report confirmed by RBC
45	Assignment of coordinate system

Table 19-11: EuroRadio track to train message types

ID	Message Name
129	Validated Train Data
130	Request for Shunting
131	Request for Supervised Maneuver
132	MA Request
133	Safe consist length information for Supervised Maneuver
136	Train Position Report
137	Request to shorten MA is granted
138	Request to shorten MA is rejected
146	Acknowledgement
147	Acknowledgement of Emergency Stop
149	Track Ahead Free Granted
150	End of Mission
153	Radio infill request
154	No compatible version supported
155	Initiation of a communication session
156	Termination of a communication session
157	Start of mission Position Report
158	Text message acknowledged by driver
159	Session Established

Table 19-12: EuroRadio train to track message types

19.7 Message Sequence and Data flows

The following diagrams provide some examples of how the messages, packets and variables are used in the exchange of data between on-board and trackside systems. The notation showing the message sequence and data flows is based on the unified modelling language (UML). Only the most important messages are shown, each one represented by their message number (Msg) and name in accordance with the ETCS SRS. Where appropriate, the relevant packets (Pk) and variables are shown in the diagrams. Some messages are optional, for instance the RBC can request a message acknowledgement by setting the M_ACK variable to '1'. Optional messages and variables are shown in italics in the diagrams.

Each element of the system is represented by a rectangle at the upper part of the diagram, e.g. the on-board subsystem or the RBC. Eurobalises are shown by a rectangle at the 'balises' line, although they are not part of the formal definition of the sequence diagrams.

The dotted vertical lines show the passage of time (from top to bottom of the diagram) for each element of the system. Time is synchronous between all the elements. The horizontal arrows representing the messages exchanged are pointing slightly downwards because the message transmission takes a finite amount of time. The time required for this is not shown to scale in the diagrams.

In addition to the UML sequence diagrams, speed/time diagrams of the trains are shown on the right. Balises are represented here by a small triangle and signals by black circles. Train movement is down the page. The dynamic speed profile represents the current and the new MA qualitatively and is not to scale.

19.8 ETCS level 1 without infill

In level 1 there is no transmission from train to track, the only information flow being from the trackside to the train. Figure 19-1 shows the basic case of a MA transmission by Eurobalise without infill, showing a train approaching a signal at red and then accelerating again when the onboard system receives the MA information which allows it to proceed.

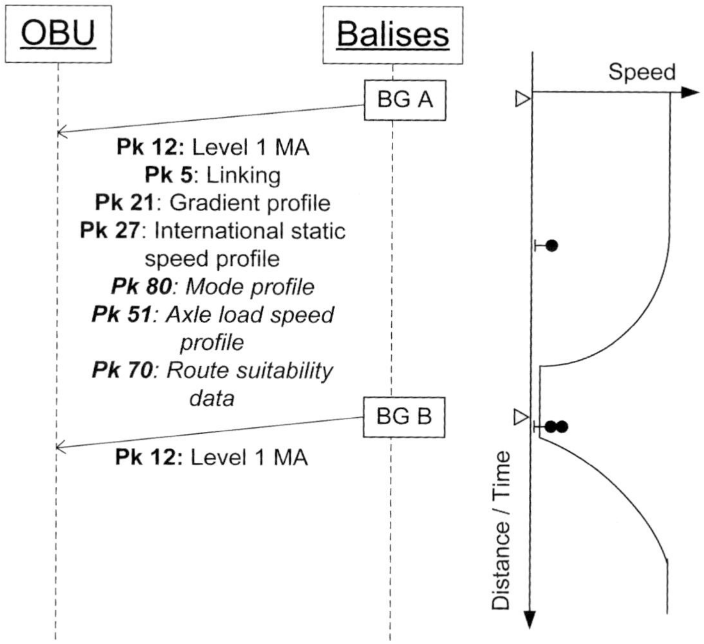

Figure 19-1: Sequence diagram for ETCS level 1 without infill

By way of an example, balise group A shows some more packets being transmitted to the train in addition to the MA packet itself. These are packets typically associated with a MA packet. Packet 27 (International Static Speed Profile), packet 21 (Gradient Profile) and packet 5 (Linking) are mandatory. The packet 80 (Mode profile), packet 51 (Axle load Speed Profile) and the packet 70 (Route Suitability Data) are examples of further packets which could be included in the data related to an MA.

Noting the speed diagram on the right in figure 19-1, the train starts to brake as it approaches the signal at red (danger). As the train passes over the associated main signal balise at the designated release speed, the on-board system receives the new MA from the balise and the driver can accelerate the train again. This is likely to be suboptimal with respect to journey time as well as energy consumption, because the signal will often clear before the train passes over the associated balise, but the on-board system is not aware of this and so will not permit the driver to accelerate until the balise is reached.

Conceivably, one solution to this problem would be to position the balise as far as possible from the main signal to ensure that trains receive up-to-date MA information before they need to start braking. However, this is inconsistent with another requirement for positioning the balise, namely that it should be as close as possible at the main signal to avoid the situation where the train has passed the balise before the new MA is transmitted (in which case the train could never receive the MA, of course).

This is a difficult optimisation problem, particularly when trains are running on the line with significantly different speeds, e.g. passenger trains with 200 km/h and freight trains with 80 km/h. Optimisation is easier in commuter or suburban networks, where all the trains travel at much the same speed. To overcome the problem of unnecessary braking on the approach to the signal, most of the current ETCS level 1 implementations use infill by Eurobalise.

19.9 ETCS level 1 with infill by Eurobalise

A typical approach for the provision of infill is to use a 'distant' balise or balise group to transmit the new MA to the on-board system before the main signal balise group is reached. Figure 19-2 shows the standard procedure using such an infill balise group (balise group BG B in the diagram). Balise groups BG A and BG C are the main signal balise groups.

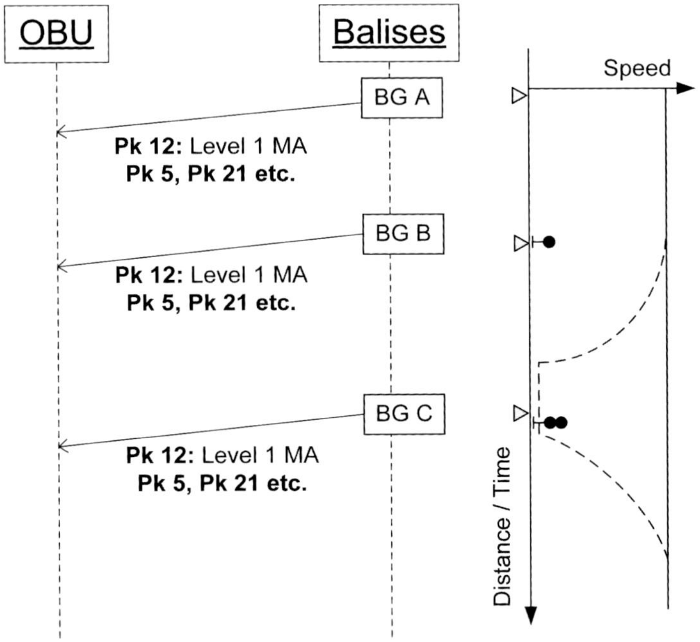

Figure 19-2: Sequence diagram for ETCS level 1 with infill by Eurobalise

Without infill the train would have to follow the speed profile shown by the dashed line, which is identical to the one in figure 19-1. With infill, balise group BG B, which is located in the vicinity of the distant (or 'approach') signal, transmits the new MA to the train if this is available when the train is passing over it. The packet structure and the content of the MA and the related data (speed profiles, track description, etc.) transmitted by an infill balise is identical to the MA transmitted by the balise at the main signal (BG C in this case).

A further infill balise is sometimes provided approximately 400 m before the main signal is reached. This enables the main signal balise to be positioned as close as possible to the signal and at the same time provide a further opportunity for the train to receive an updated MA in the event that it could not do so when it passed over the 'distant' balise (BG B).

19.10 ETCS level 2

In ETCS level 2 a radio connection is used to transmit MA information, and a trackside Radio Block Centre (RBC) controls the train movement, so sending an updated MA to the train is no longer restricted to specific locations or areas. The RBC could, for instance, send a new MA for a route that has been set further ahead just before the train reaches the beginning of the braking curve for stopping at the End of Authority location associated with the current MA.

The permanent availability of the full bi-directional radio connection as well as a high frequency update of the train location to the RBC enables the latter to control the train movement with a greater flexibility than is possible in level 1. An information exchange between trackside and the on-board system to facilitate this can be initiated whenever required.

Figure 19-3 shows the data exchange for the MA transmission in level 2, starting with the initiation of a communication session on entry into a level 2 area.

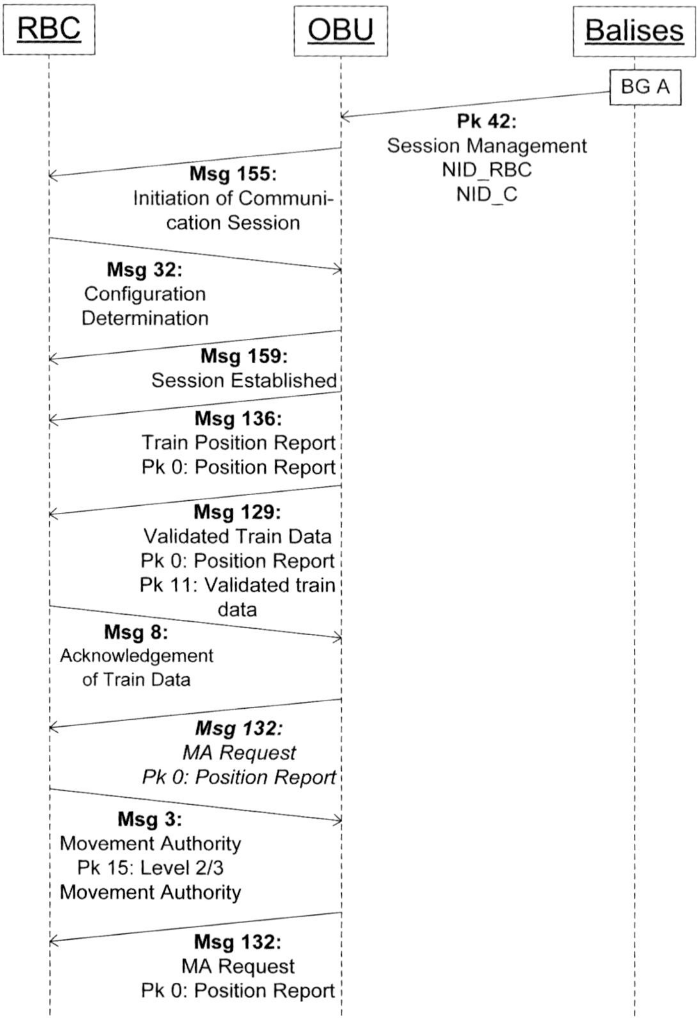

Figure 19-3: Sequence diagram for ETCS level 2

The on-board system needs to know the numerical identifier of the country (NID_C) as well as the numerical identifier of the RBC (NID_RBC). They can be transmitted by a balise group (BG A in the example) before the level 2 area is reached. This is typically the case when the train is performing a transition from a level 1 area into a level 2 area. The level 2 communication session is then initiated by the on-board system using the standard sequence of message 155 (Initiation of communication session), followed by message 32 (RBC/RIU System Version) and then message 159 (Session Established). The session can also be initiated from the trackside sending the message 38 (Acknowledgement of session establishment) from the trackside to the on-board system.

Having established the level 2 communication session, the first set of information sent from the on-board system to the RBC is message 129 (Validated Train Data), which includes packet 0 (Position Report) and packet 11 (Validated train data). They are used to inform the RBC about various items of train data such as length, axle load, loading gauge and maximum speed of the train, as well as the current position and time of the on-board clock. The RBC acknowledges the train data by sending message 8 (Acknowledgement of Train Data). This message contains a peculiarity, namely that the variable T_TIME is used twice. The first one is the estimated on-board time at the moment of sending. The second one is the T_TIME of the previous message 129 from the train to the RBC. This mechanism is used to synchronise the on-board and RBC times and to supervise the sequence of the radio messages.

Once this initial exchange has taken place, the train can send a MA request (message 132) including a position report (packet 0 or packet 1). The RBC sends a new MA if it is able to do so, fulfilling the requirements mentioned above. This uses the message 3 (Movement Authority) which contains packet 15 (Level 2 Movement Authority) and the other packets needed for a complete MA. The MA request will typically be generated by the on-board system when it identifies that the train is approaching the location where the braking curve begins for the end of the current MA. Alternatively, the RBC may send a movement authority to the train without it being requested by the on-board system, for instance when the preceding train has moved and the train's current MA can therefore be extended.

Level 2 offers a higher degree of flexibility with respect to the block section lengths compared with level 1, as it is no longer constrained by lineside signal locations and braking characteristics. Level 2 with short block lengths and no lineside signals can achieve higher track capacity than typical lineside signalling systems.

If the on-board equipment includes a train integrity proving system, this status will be reported in the variable Q_INTEGRITY as part of the regular position reports to the RBC, and this can be used by the RBC to determine the location (front and back) of the train independently of conventional lineside train detection systems. This allows the MA for a following train to be generated on a 'moving block' principle, which yields further improvements in track capacity. In this scenario the frequency of the train position reports from the on-board system becomes a critical determining factor for capacity.

References

SUBSET 026-7: System Requirements Specification Chapter 7 ERTMS/ETCS Language

SUBSET 026-8: System Requirements Specification Chapter 8 Messages

Chapter 20: ETCS On-board Interfaces

20.1 Generic considerations for on-board architecture

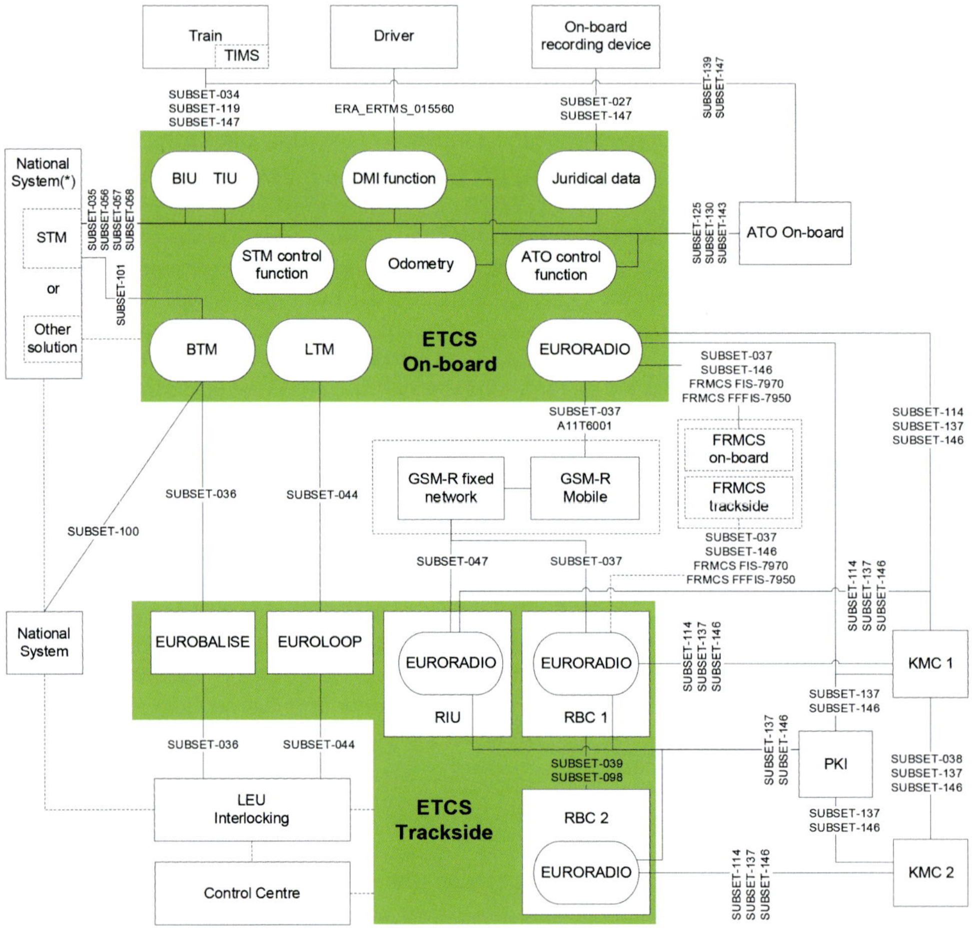

(*) Depending on its functionality and the desired configuration, the national system can be addressed either via an STM using the standard interface or via another national solution

Figure 20-1: ERTMS/ETCS reference architecture (from SUBSET-026-2)

The interfaces between the ETCS on-board system to the driver via the DMI and to the ETCS trackside via balises and radio are described in other chapters. In this chapter we describe the interfaces of the ETCS on-board to other systems on the train:

- Train interface (brakes, traction, etc.)
- On-board recording device (formerly known as JRU)
- National (Class B) automatic train protection systems
- Automatic Train Operation (ATO) on-board

20.2 Train interface

The train interface is the boundary between the ETCS on-board and the railway vehicle on which it is installed. The ETCS on-board receives certain information from vehicle systems (inputs) and distributes information to the vehicle systems (outputs).

The interface is described at a functional level in UNISIG SUBSET-034 ("Train Interface FIS"). From ERTMS Baseline 4 onwards the technical and physical details are also standardised in SUBSET-119 ("Train Interface FFFIS"). The aim of this is to minimise the number of interfaces and components needed for the integration of ETCS onto a vehicle and to avoid having to make changes to a vehicle when upgrading the on-board ETCS – and vice versa. Safety requirements for the train interface are specified in SUBSET-120 ("FFFIS TI – Safety-related Requirements").

The train interface consists of a hard-wired interface and a serial data interface. The serial interface uses one of the options specified in the IEC 61375 family of standards for train communication networks, namely MVB (Multifunction Vehicle Bus), CAN (Controller Area Network) or ECN (Ethernet Consist Network). The hard-wired interface is mandatory for three critical functions, namely emergency brake, ETCS isolation and traction cut-off. For all other functions the serial interface is mandatory, although in most cases a hard-wired alternative option is also allowed. This option may be necessary for retrofitting ETCS to an older vehicle that is not fitted with a train communication network.

Category	Function	Source	Hard-wired interface	Serial interface
ETCS mode control	Sleeping	Train	O	M
	Passive Shunting	Train	O	M
	Non-Leading	Train	O	M
	Isolation (of ETCS)	ETCS	M	–
	Automatic Driving	ETCS	O	M
	Remote Shunting	ETCS	O	M
Control of brakes	Service Brake Command	ETCS	O	M
	Brake Pressure	Train	–	M
	Emergency Brake Command	ETCS	M	M
	Regenerative Brake Inhibit	ETCS	–	M
	Magnetic Shoe Brake Inhibit	ETCS	–	M
	Eddy Current Brakes for Service Brake Inhibit	ETCS	–	M
	Eddy Current Brakes for Emergency Brake Inhibit	ETCS	–	M
	Special Brake Inhibit – STM Orders	ETCS	O	M
	Special Brake Status	Train	O	M
	Additional Brake Status	Train	O	M
Control of train functions	Change of Traction System	ETCS	–	M
	Powerless Section with Pantograph to be Lowered – Trackside Orders	ETCS	–	M
	Pantograph – STM Orders	ETCS	O	M
	Air Tightness – Trackside Orders	ETCS	–	M
	Air Tightness – STM Orders	ETCS	O	M
	Station Platform	ETCS	–	M
	Powerless Section with Main Power Switch to be Switched Off – Trackside Orders	ETCS	–	M
	Main Power Switch – STM Orders	ETCS	O	M
	Change of Allowed Current Consumption	ETCS	–	M
	Engine Orientation in Supervised Manoeuvre	ETCS	O	M
	Traction Cut-Off	ETCS	M	M
Train status information	Cab Status	Train	O	M
	Direction Controller	Train	O	M
	Train Integrity	Train	O	M
	Traction Status (only for STM)	Train	O	M
	Set Speed (for DMI indication)	Train	–	M

▶

Category	Function	Source	Hard-wired interface	Serial interface
Train data	Type of Train Data Entry	Train	O	M
	Overall Consist Length Information	Train	–	M
	Train Data Information	Train	O	O
	Train Running Number	Train	–	M
National system status	National System Isolation	Train	O	M

Table 20-1: List of train interface functions

The allocation of functions to the hard-wired and serial interfaces is shown in table 20-1. For each specific on-board subsystem application, a subset of the interfaces will be relevant; only Isolation, Emergency brake and Cab status are mandatory in all applications.

Note that mandatory (M) in the table also indicates that the function must be provided in the ETCS on-board system for it to be approved as an interoperable constituent that is suitable for use in any type of train.

The hard-wired interface works by using isolated outputs (voltage free contacts) on the sending system which switch an electrical current that is sourced from the receiving system. The standard allows for a range of voltages from 24 V to 110 V. The voltage for outputs from the ETCS on-board is determined by the vehicle system and the voltage for the inputs by the ETCS on-board.

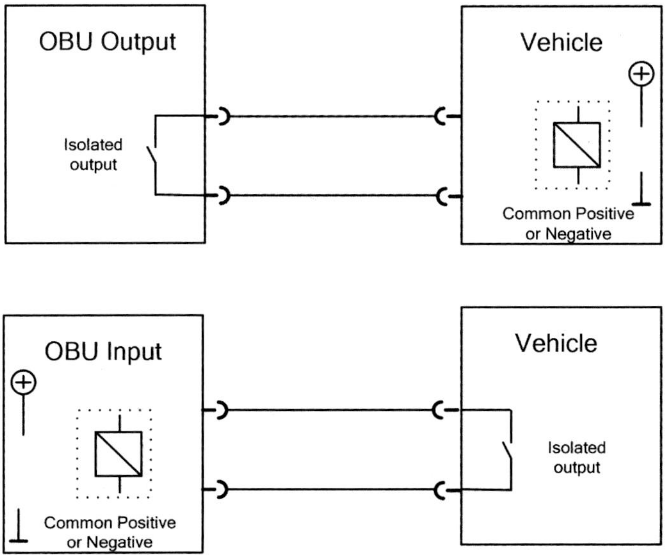

Figure 20-2: Hard-wired interface arrangements (from SUBSET-119)

The emergency brake output is duplicated for safety, either by means of two hard-wired outputs, or by one hard-wired output and a message via the serial interface. To achieve the required level of safety it is necessary to undertake a regular test that both channels are functioning correctly. This is typically undertaken as part of the ETCS system startup process when a cab is 'opened', or as a minimum on a daily basis.

For the serial interface the functions are grouped into 24-byte packets which are then embedded in messages in a format appropriate for the type of vehicle bus in use. There are three packets of data for inputs from the train to the ETCS on-board, and seven packets for outputs from the on-board to the train.

20.3 On-board data recording device

An ETCS fitted train will have an on-board data recording device which stores information about the operation of the train. This is used for subsequent analysis to review performance or investigate an accident. The functional, performance and data integrity requirements for the data recording device are specified in the Technical Specification for Interoperability relating to Locomotives and Passenger Rolling Stock (TSI LOC & PAS), which in turn references the standard EN 62625-1:2013 (Electronic railway equipment – On board driving data recording system – Part 1: System specification).

The functional interface between the ETCS on-board and the data recording device is specified in SUBSET-027 (FIS for Juridical Recording). Earlier versions of the ETCS standards specified a dedicated juridical recording unit (JRU) as part of the CCS on-board subsystem, but the data recording for ETCS is now integrated with recording of other train functions.

The Juridical Recording FIS defines the format and content of the data messages sent by the ETCS on-board, as well as the list of on-board events that trigger the transmission of the related messages. For each event the ETCS on-board equipment registers the date and time using Universal Time Co-ordinated (UTC), the train location and speed and the ETCS system version, level and mode at the time the event occurred.

Some of the events which trigger a data message to the data recording device are:

- Changes of ETCS level or mode
- Messages received and sent on any of the interfaces of the ETCS on-board system
- Updates to the information displayed on the DMI
- Input of information by the driver into the DMI
- 5 seconds elapsed since the last trigger event occurred

The TSI CCS (Baseline 4) does not specify the physical interface from the ETCS on-board to the data recording device, but it is envisaged that in future this will use the standard for network technology that is specified for communications between ETCS, ATO and FRMCS. This is defined in SUBSET-147 (FFFIS part: CCS Consist Network Communication Layers) and is based on Ethernet technology.

20.4 Interface with other train protection systems

Even though ETCS is the European standard train protection system, existing national train control systems will still be used for many years. It is very important to ensure that ETCS-equipped trains can still run on lines equipped with existing systems. This is recognised in the

TSI CCS which designates these as 'Class B' systems, which may continue to be installed and used during the migration to ETCS.

The co-existence of both systems (ETCS and Class B) in a single vehicle requires at least that the transitions between them are handled in a safe and efficient way. When the train is running under the supervision of the Class B system, the on-board ETCS is operating in what is known as Level NTC (national train control).

There are several options for implementing ETCS alongside a national ATP system on a train. These are illustrated in figure 20-3. The Class B train protection system and the ETCS on-board can be completely independent systems, each with its own interface to the driver and other train systems (Option 1). The only interface between the systems in this scenario is the enabling/disabling of the Class B system at mode transitions. The disadvantage of this solution is that several functions need to be duplicated on the train, for example the DMI, odometry, the JRU and train interfaces. This can increase costs and give rise to installation problems. It may also cause ergonomic and human factors concerns, as drivers must split their attention between two different displays in the cab.

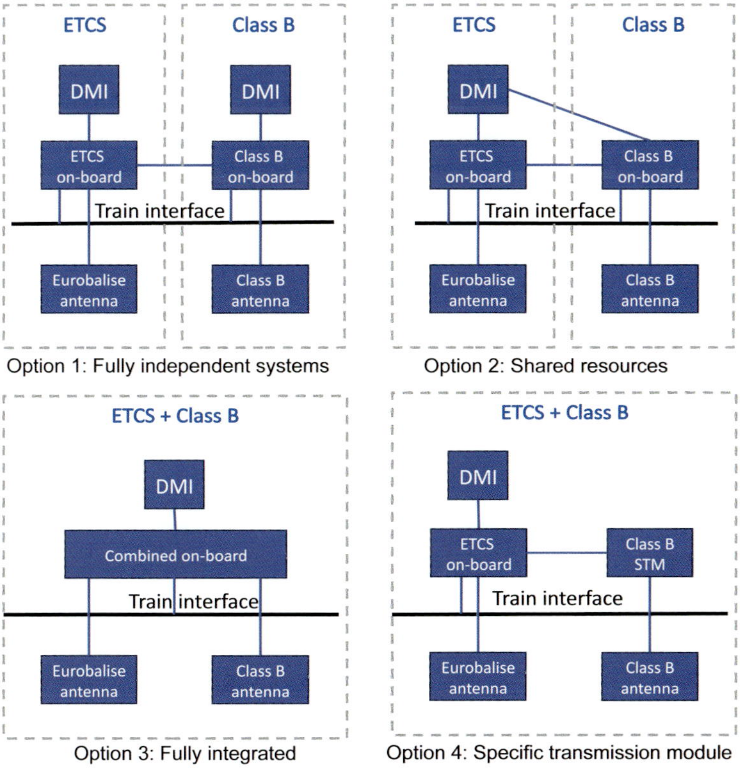

Figure 20-3: Options for interface between ETCS and Class B on-board systems

To mitigate these issues, it is possible to share some resources between the two systems. In the example shown (Option 2) the DMI is shared between the ETCS and Class B systems, but this could also apply to other components such as odometry, juridical recorder and train interfaces. The interface may be direct from the Class B on-board to the shared component, or via the ETCS on-board system. This reduces the duplication of equipment between the ETCS and Class B systems, and in the case of a shared DMI also reduces the ergonomic and human factors problems posed by having multiple DMIs in the cab. However, sharing equipment requires adaptation of both the ETCS interfaces and Class B systems.

Another possibility (Option 3) is that the Class B train protection system and the ETCS functionality are implemented in a single integrated system. This is sometimes referred to as a 'bi-standard' on-board system. It can be a cost-effective and efficient solution that eliminates all duplication of equipment but requires bespoke development which can only be undertaken by a supplier with the knowledge and intellectual property rights for the Class B system.

Option 4 uses the ETCS 'specific transmission module' (STM) concept, which has been developed to enable the functionality of the Class B system to be implemented using the resources of the ETCS on-board system. An STM is a device that provides the Class B system functionality and which interfaces with the ETCS on-board equipment to access components such as the DMI, JRU, odometry, the train interface, braking systems, etc. By using an STM, the functionality of a Class B system can be implemented with a minimum of additional hardware on the train – usually all that is required apart from the STM itself is an additional track to train interface antenna, hence the name 'specific transmission module'. The possibility of installing multiple STMs linked to the on-board ETCS by a common data bus is particularly useful for international trains that are required to interface with several different Class B systems.

If necessary, the STM solution can be combined with the other approaches, for instance the Elizabeth line trains in UK operate under ETCS level 2 and the Class B system TPWS on the national network, and a proprietary CBTC (communications-based train control system) on metro tracks in London. In this solution the TPWS is implemented as a STM while the CBTC is a separate system with some shared resources.

20.5 Specific Transmission Module (STM)

Development of a STM is a significant investment which requires knowledge and intellectual property rights for the Class B system. This may require sponsorship of the development by a national government, infrastructure manager or railway undertaking that uses the Class B system, but once the STM is available for a specific Class B system it can be integrated with any supplier's ETCS on-board equipment. This solution can also help manage obsolescence issues for the Class B system, as new trains can be fitted with the ETCS + STM solution instead of the obsolete Class B on-board, even if the trains run only on Class B lines initially.

To enable this concept to be realised, it was important to standardise the interface between an STM and the on-board ETCS, to allow interoperability between suppliers. This is necessary because in many cases there will only be one supplier of an STM for a specific national system, either because the market size is limited or because of intellectual property constraints. The STM interface is based on the industrial PROFIBUS standard and five UNISIG specifications:

- SUBSET-035 Specific Transmission Module FFFIS
- SUBSET-056 STM FFIS Safe Time layer
- SUBSET-057 STM FFIS Safe Link Layer
- SUBSET-058 FFFIS STM Application Layer
- SUBSET-059 Performance Requirements for STMs

The physical layer of the STM-ETCS connection is an RS-485 serial interface operating at a baud rate of 1500 kbps, and the default physical medium is a twisted pair shielded copper cable. Multiple STMs can be connected to one physical bus and the protocols allow for this to be duplicated for availability.

The application layer is specified as a 'language' in a similar manner to the track to train interface. Messages to and from the STM are made up of packets and variables that allow the exchange of data relating to the various elements of the ETCS on-board system that can interact with the Class B functions in the STM. These include:

- STM state report and system version information
- STM control
- Odometer
- Driver machine interface
- Train interface
- Data recording

When running in level NTC, the responsibility for the train supervision is transferred by the ETCS on-board subsystem to the corresponding Class B functionality implemented in the STM, which uses the information it receives directly from the Class B antenna together with the information received from the ETCS. It interacts with the driver via the ETCS DMI and with train systems and brakes via the ETCS train interface. The ETCS on-board subsystem remains active, but mainly to monitor trackside information related to the level transition process and to supervise the potential failure of the STM.

The transitions from and to areas covered by the corresponding Class B system are processed as level transitions to or from Level NTC and are controlled by the ETCS on-board subsystem. During the level transitions, the ETCS on-board subsystem controls the activation and de-activation of the corresponding STM. On the approach to the border between the two levels the ETCS trackside subsystem delivers the necessary information to perform the level transition via balise groups or via the RBC.

A typical example of an STM is one for the UK national system TPWS, which provides warning, speed trap and train stop functionality at lineside signals and speed restrictions. The TPWS STM has an on-board antenna that detects the magnets and inductive loops on the track, and it interfaces with the on-board ETCS to apply the train brakes and display the traditional yellow and black 'sunflower' indication and warning sounds on the ETCS DMI.

A special case exists for Class B systems that use a similar remote powering technology to that of the Eurobalise. These are known as the 'KER' systems, KVB, EBICab, and RSDD. It is possible to design an ETCS BTM that is capable of reading both types of balise and a dedicated interface (SUBSET-101 Interface 'K' Specification) has been specified to transfer data received via the airgap from a KER balise directly from this type of BTM to the STM for one of these Class B systems.

An STM may also make use of packet 44 to allow it to communicate via the Eurobalise or EuroRadio transmission systems. This has allowed modernising of some existing national systems to use Eurobalise for the track to train interface in place of an older technology. This may be motivated by obsolescence of the legacy system equipment or to facilitate the later migration to ETCS. Examples where this has been done include the following countries and national train protection systems:

- Italy – SCMT
- Belgium – TBL1+
- Switzerland – EuroZUB

This solution allows upgrading of existing trains with an ETCS BTM in place of the original national system antenna (sometimes referred to as a 'reverse STM') whilst retaining the other train interfaces of the original system This can be significantly less costly than a full ETCS installation and is suitable for trains with a limited remaining lifetime that are unlikely to need to run on an ETCS fitted track. For ETCS fitted trains, the STM required to work with this type of national system can be simply implemented in the form of a software module, as there is no need for an additional external interface. Migration to ETCS level 1 can be achieved by programming the trackside balises with ETCS level 1 information alongside the packet 44 national system information and transmitting both to the train. This then allows any ETCS train to operate without an STM, while other trains can continue to run using the national system information via packet 44.

20.6 ATO interface

With the extension of scope of ERTMS to include automatic train operation (ATO), introduced in the 2023 TSI CCS as described in chapter 13, there is a requirement for a Baseline 4 ETCS on-board to exchange information with the ATO on-board systems. The information to be exchanged is defined in SUBSET-130 (ERTMS/ATO – ATO-OB / ETCS-OB FFFIS Application Layer).

The FFFIS Application Layer specifies five packets of information to be sent from the ATO on-board to the ETCS on-board, including:

- Request for the ETCS on-board to enter AD (Automatic Driving) mode
- ATO information to be displayed on the DMI
- Requests for ATO data entry by the driver
- View of entered data to be displayed to the driver

There are also seven packets specified to be sent from the ETCS on-board to the ATO on-board, including:

- Static information about the train, e.g. brake performance
- Dynamic information about the train, e.g. location and speed
- Driver inputs and data entry
- Safe emergency brake deceleration values

Some of the packets are sent on a cyclic basis at intervals of 100-1000 ms and others are event based.

Lower-level details of the interface are specified in SUBSET-143 (ERTMS/ATO Interface Specification – Communication Layers for On-board Communication) which in turn references IEC 61375-3-4:2014 Electronic railway equipment – Train communication network (TCN) – Part 3-4: Ethernet Consist Network (ECN).

References

SUBSET-034: Train Interface FIS

SUBSET-027: FIS for Juridical Recording

SUBSET-035: Specific Transmission Module FFFIS

SUBSET-056: STM FFIS Safe Time layer

SUBSET-057: STM FFIS Safe Link Layer

SUBSET-058: FFFIS STM Application Layer

SUBSET-059: Performance Requirements for STMs

SUBSET-101: Interface 'K' Specification

SUBSET-119: Train Interface FFFIS

SUBSET-120: FFFIS TI – Safety-related Requirements

SUBSET-130: ERTMS/ATO – ATO-OB / ETCS-OB FFFIS Application Layer

SUBSET-143: ERTMS/ATO Interface Specification – Communication Layers for On-board Communication

SUBSET-147: FFFIS part: CCS Consist Network Communication Layers

CENELEC EN 50170-2:1996: General purpose field communication system (PROFIBUS)

IEC 61375-3-4:2014: Electronic railway equipment – Train communication network (TCN) – Part 3-4: Ethernet Consist Network (ECN)

ERA/TD/2011-11: List of CCS Class B systems

Chapter 21: Driver Machine Interface

21.1 Introduction

The driver machine interface (DMI) in the train cab is one of the most distinctive and visible features of ETCS. The DMI format was the result of an ergonomics study undertaken by the UIC in the early 1990s, with the concept being validated in a cab simulator using train drivers from several countries.

Both the train and the driver are the responsibility of the Railway Undertaking (RU), and therefore the DMI did not require to be standardised to achieve technical interoperability. Nevertheless, the specification for the DMI has been made mandatory, so that drivers operating different trains and working for different RUs always interact with a consistent DMI regardless of which supplier's products are fitted. This facilitates cross-acceptance of equipment and reduces the potential for human errors and the driver training requirement.

The purpose of the DMI is to allow the driver to:

- enter data about the train into the system and validate it at start of mission;
- display all the information that the signalling system provides for safe and efficient driving of the train;
- allow for driver acknowledgement of information display when required.

Figure 21-1: Typical DMI installation in cab, Class 102 train, Madrid (Siemens)

The DMI hardware is a colour display screen with driver input facilities, normally positioned directly in front of the driver in the train cab. The specification allows for two versions, depending on the technology used for driver input. It can be either a 'touch screen' where the driver interacts by directly touching areas of the screen that are configured to act as buttons, or one with separate hardware buttons alongside the screen, with the function of each button being indicated by an adjacent 'soft key' on the screen. The DMI also incorporates an audio output to alert the driver to situations where some action may be required.

21.2 Screen layout

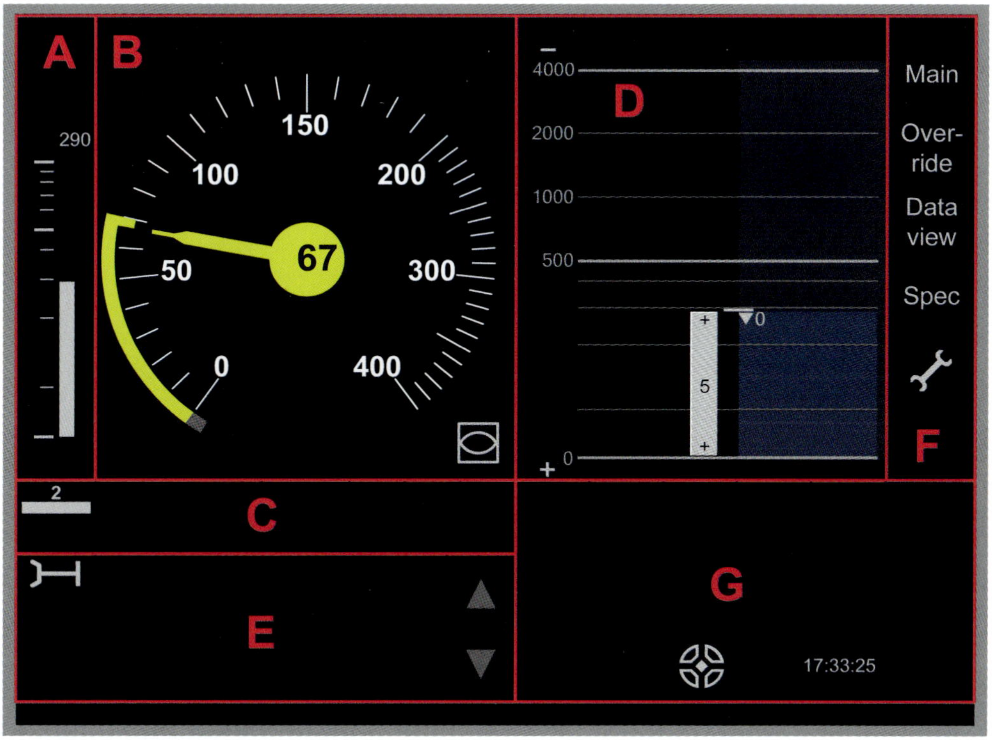

Figure 21-2: ETCS DMI screen areas – touch screen version

The DMI specification sets out in detail how the screen is to be divided into areas and sub-areas for the display of symbols, numbers and text. There are seven main areas, with the following functions:

A – Brake details area

This is a strip to the left of the main speedometer which shows information to assist the driver in braking to a new target speed or limit of authority, showing:

- distance to target (a vertical bar and number of metres)
- time to change in ETCS supervision status (a square that increases in size)
- low adhesion symbol

B – Speed control area

This is the main circular speedometer display, showing:

- current speed of the train (position of pointer and the number in the centre)
- permissible speed (a 'hook' on the speedometer dial)
- target speed (change in colour on the speedometer dial)
- release speed (change in colour on the speedometer dial)
- ETCS mode
- current orders and announcements of track conditions

C – ETCS level and mode information

This is the strip below the speedometer, showing:

- ETCS level
- change of mode acknowledgement
- override
- tunnel stopping area
- reversing permitted

D – Planning area

This area gives the driver advance information about the route ahead of the train, showing:

- gradient profile
- permissible speed profile
- location of speed profile discontinuities and end of movement authority
- location of orders and announcements of track conditions
- indication marker

This area is also used for data entry and driver input of Track Ahead Free information.

E – Text messages

This area displays text messages to the driver and an indication of a safe radio connection in level 2. The most recent message is at the top, and the display can be scrolled to view older messages. Messages shown include:

- system status messages
- text messages from the trackside
- safe radio data connection

F – Driver input

This area contains buttons that allow the driver to display sub-level windows for data entry purposes:

- Main: start, driver ID, train data, level, training running number, shunting, non-leading, radio data.
- Override: override end of authority.
- Special: adhesion, staff responsible speed/distance, train integrity.
- Settings: language, volume, brightness, system version, set/remove virtual balise cover.

G – Miscellaneous

This area displays the local time as known to the ETCS on-board, and other optional items of information such as:

- geographical position
- ATO information

Some of the cells in areas C, E and G are not allocated to an ETCS function, so they are available for other purposes, such as advisory speeds as calculated by an automatic train operation (ATO) system. In Level NTC, some areas can be reprogrammed to provide indications specific to the national train control system – area D is often used for this purpose.

21.3 Use of colour

Colours on the DMI for symbols and text denote the importance and priority of information displayed, and the urgency of action that may be required by the driver, as indicated in table 21-1.

WHITE OR GREY	YELLOW	ORANGE	RED
Usual situation	Usual situation	Unusual situation	Unusual situation
Expected	Expected	Maybe not expected	Generally not expected
No danger at all	Not dangerous YET	BECOMING dangerous	DANGEROUS situation
No action required by driver	You still have some time to take action.	You CAN STILL PREVENT	TOO LATE TO PREVENT
Use the information when you like.	You can do other things first but do not forget.	Take action NOW!	System has INTERVENED
	If you do not react, this might turn orange.	If you do not react, this might turn red.	You may need to acknowledge to regain control.

Table 21-1: Use of colour on the DMI

Figure 21-3 illustrates the specific example of the speed indication and how the colour of the speedometer pointer and the circumference of the dial is based on the current speed of the train with respect to the permitted speed and service brake supervision speed limits applicable at the current location within the movement authority.

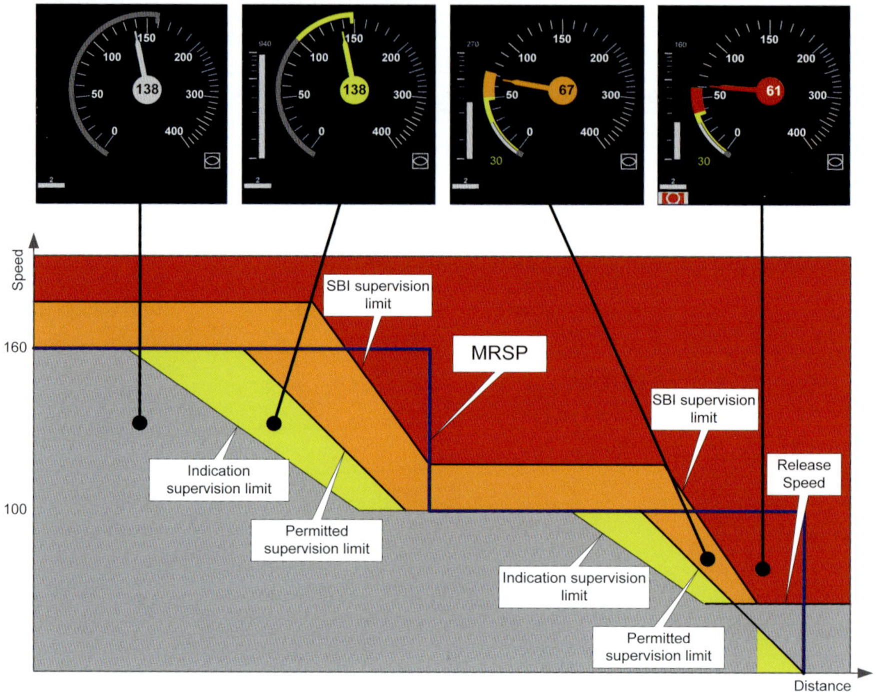

Figure 21-3: Colours linking to supervision limits

21.4 DMI symbols

Symbols are defined in the DMI specification to minimise the amount of text that needs to be translated into different languages for international trains. The use of colour in the symbols follows the rules described above, and flashing symbols are used to draw attention to a condition that requires acknowledgement or another action by the driver. Some examples are shown in figure 21-4:

MODES (area B)	Full Supervision 	On Sight  Flashing yellow when acknowledgement required	Staff Responsible  Flashing yellow and larger when acknowledgement required	Trip  Flashing yellow and black background when acknowledgement becomes possible
LEVELS (area C)	Level 1 	Announcement of Level 0 	Level 2 	Level NTC  Text NTC is replaced with abbreviation for national system
TRACK CONDITIONS (area B)	Lower pantograph 	Radio hole 	Change of traction system announcement 	Tunnel stopping area 

Figure 21-4: Examples of DMI symbols

21.5 Data entry

When driver input is required (other than a simple acknowledgement), one of a number of sub-level windows will be displayed. These are of various types:

- Menu windows
- Data entry windows
- Data validation windows
- Data view windows

On a touch screen DMI the window will display 'buttons' to be touched directly by the driver, while on a soft key DMI there are numbered options that refer to the hardware buttons around the screen. Figure 21-5 shows both the touch screen (top) and soft key (bottom) versions for the main menu used to select the type of data to be entered (left) and, as a specific example, for train running number entry (right).

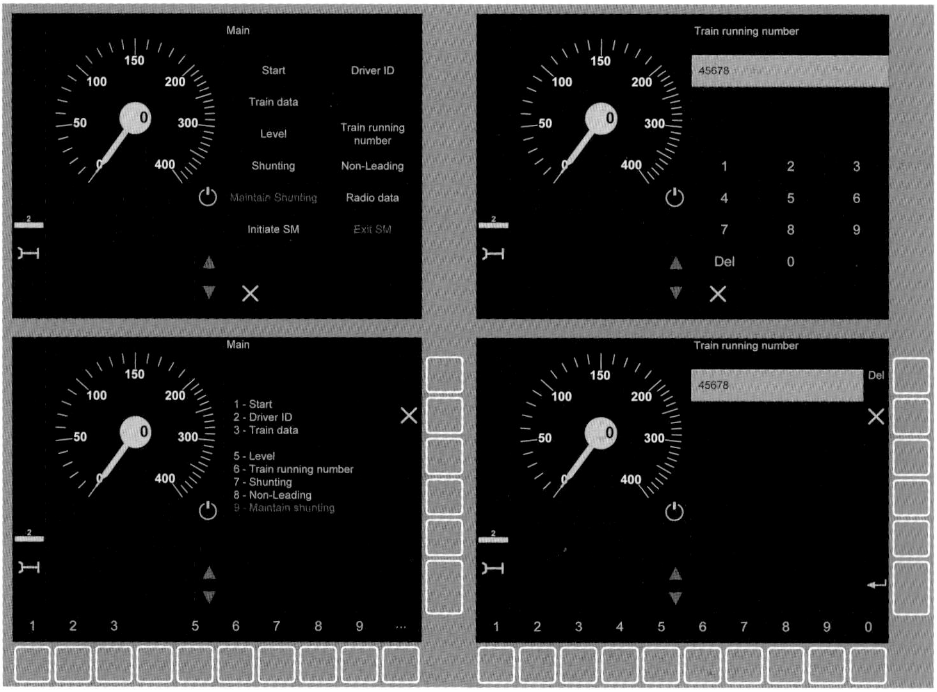

Figure 21-5: Main menu (left) and train running number data entry (right)

21.6 Safety and availability

The DMI is a safety related component of the system, and a functional safety analysis has been undertaken to determine the safety integrity level (SIL) that is necessary for each function to achieve a tolerable hazard rate. This determines the software development and verification techniques that must be followed for DMI development in order to comply with CENELEC standards. Many of the functions have a low impact on safety because the consequences of an incorrect display are mitigated elsewhere in the system. For example, in FS mode if the permissible speed or actual speed is displayed incorrectly on the DMI, the ETCS on-board will still intervene to apply the brakes if necessary. Where data entry by the driver is required, there is a confirmation process where the entered information is displayed back to the driver to be checked before it is acted upon. The most critical functions are those where the driver has to rely on DMI information in a degraded mode of operation, or where the DMI is used to isolate or disable functionality. SILs assigned to the DMI functionality are as follows:

- SIL 2 – Current speed indication, isolation command, virtual balise cover.
- SIL 1 – Fixed text message indication, mode indication, acknowledgement of intervention or change of mode.

The reliability of the DMI is also critical, as it provides the primary indication of train speed to the driver, even when the train is not operating in an ETCS mode. For this reason, most applications provide a backup DMI or a secondary speedometer elsewhere in the train cab so that the train can still be driven in the event of a total failure of the primary DMI. Another solution is to provide a split DMI, with two portrait format screens side by side and a facility to move the most critical information between screens in the event of a failure.

21.7 Use of DMI with a national train control system

When a national train control system is integrated with the ETCS on-board in the form of a STM, the controls and indications required for the national display can be integrated into the ETCS DMI. The STM FFFIS application layer specification defines a language of packets and variables that allow the STM to communicate with the ETCS on-board to send and receive information to and from the DMI. Two options are available for implementing this facility, namely the 'customisable' and the 'unified' DMI services.

A customisable DMI is one which includes the specific functionality for the national train control system that the trains will be using. For instance, special symbols can be displayed that are not included in the ETCS DMI specification. This option provides maximum flexibility for changing the appearance of the screen to operate in level NTC, but it requires a special version of DMI software which is compatible with the STM that is being used.

A unified DMI is one which includes only generic functionality that allows national train control system functions to be implemented either using elements of the standard ETCS DMI interface, or by simple blocks of colour and text. This option avoids the need to develop a special version of DMI software for each type of STM, but the disadvantage for drivers is that the display may look quite different to that of the national system with which they are familiar.

References

ERA_ERTMS_015560: ETCS Driver Machine Interface

SUBSET-028: FFFIS STM Application Layer

SUBSET-118: Functional Safety Analysis of ETCS DMI for ETCS Auxiliary Hazard

ERA Generic ETCS Driver's Handbook

Chapter 22: Interface between Radio Block Centres

22.1 Introduction

The size of the control area with which an RBC can deal is limited by its processing and interfacing capacity. In addition, there are other reasons, such as country borders, and the borders between sections of line fitted with ETCS equipment supplied by different manufacturers, that may impose limits on the size of the RBC control area. At all these locations, each RBC needs to communicate with its neighbouring RBC.

Originally the communication between RBCs was not harmonised, except for the information exchanged between trackside and ETCS on-board for the purpose of an RBC-RBC handover. However, the variety of different solutions developed by the suppliers soon became an issue when interfacing RBCs. To address this problem, specifications for a harmonised RBC-RBC interface were developed. These are:

- SUBSET-039: FIS for the RBC/RBC Handover
- SUBSET-089: RBC-RBC Safe Communication Interface

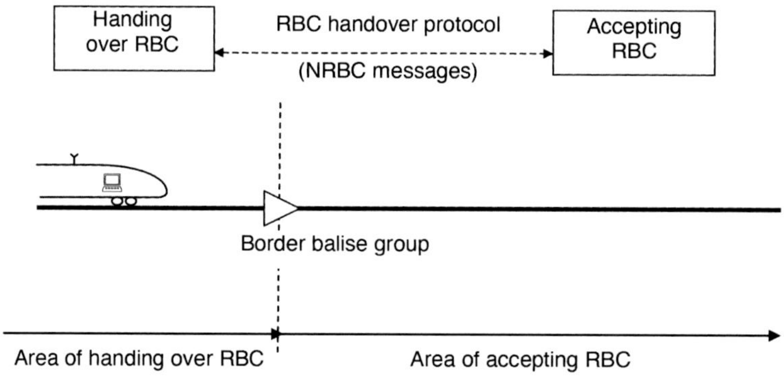

Figure 22-1: RBC-RBC handover concept (from SUBSET-039)

Even with these specifications, the interface cannot be fully harmonised. A considerable amount of configuration is necessary to allow both RBCs to communicate smoothly and the transition of the train from one RBC to another is seamless, so that in normal situations no operational restrictions exist (the train can cross the transition at line speed) and the transition is not apparent to the driver.

To ensure an efficient and smooth handover it is necessary for the train to be able to manage two simultaneous communication sessions, one with each RBC. However, the handover process must also be possible for trains having only one session available, for instance in the event of a failure of one on-board mobile radio.

In some cases, the RBC-RBC border corresponds to a border between mobile network providers. This means that the opening of the second (accepting) session requires registration on a new mobile network in order to avoid disruption of communications when passing over the border.

The RBC-RBC handover process requires that the request to establish a second (accepting) communication session is initiated when the movement authority of the train reaches the

RBC-RBC border. This means that when trains require long movement authorities (particularly relevant to high speed trains), the handover process must be initiated a considerable distance before the border. Depending on the track topology, this requires the coverage of both GSM-R networks to overlap extensively, which is usually difficult to achieve.

22.2 Principles of RBC-RBC handover

The three main principles for RBC-RBC handover from a 'handing over' (HOV) RBC to an 'accepting' (ACC) RBC are:

1) The communication between the ETCS on-board and the HOV RBC is maintained until the full length of the train (minimum safe rear end) has passed the border between the RBCs. The reason for this is that in ETCS level 2 with on-board train detection, the RBC needs to know when the train has completely left its area. In ETCS level 2 using trackside train detection (axle counters and track circuits), the interlocking provides this information to the RBC.
2) If the train has two ETCS mobiles available, and the accepting RBC is already in communication with the ETCS on-board, responsibility for the train transfers from the HOV RBC to the ACC RBC when the 'maximum safe front end' of the train passes the border between the RBCs.
3) If the ETCS on-board has only one ETCS mobile available (because the other one has failed), the transition of responsibility is delayed until the first principle is fulfilled, the communication with the HOV RBC has been terminated, and the communication with the ACC RBC has been established. So there is a gap in time in which there is no radio communication with the train.

An interface is provided between neighbouring RBCs by which the ACC RBC can forward route related information (RRI), using the same data packets that are used for the track/ train communication, with location information referenced to the border balise group. This information will typically include:

- Movement authority information
- Balise linking information
- Track description information such as static speed profiles, temporary speed restrictions, level crossing locations

This enables the HOV RBC to extend the movement authority information beyond the RBC-RBC border. For this purpose, the HOV RBC has to combine the RRI information from the neighbouring RBC with the information about its own area. This implies that the distances in the RRI information have to be modified to refer also to the current last relevant balise group (LRBG) of the ETCS on-board.

The exception to this handover process is where an Emergency Stop message is issued by the ACC RBC. Emergency stop messages are not exchanged between both RBCs as it would require a difficult handshake between the RBCs. Instead, an updated RRI message is sent to the HOV RBC to shorten the movement authority beyond the border.

With the introduction of system version management (part of ETCS Baseline 3 onwards), another difficulty had to be solved. When coupling two RBCs that have different system versions, the RBC with the higher version must be able to process the RBC/RBC messages in the language which corresponds to the system version of the other RBC.

Degraded situations may lead to the train being stopped at the border between adjacent RBCs. In these cases, operational rules will usually require the driver to use the ETCS override function in co-ordination with the traffic control centre. This will enable the train to pass the border without generating an emergency brake application (via the train trip function). After passing the border, the process for recommencing normal operation is similar to the train awakening procedure in the area of the ACC RBC.

22.3 The steps of an RBC-RBC handover

Step 1 (Pre-announcement)

In this initial step, the HOV RBC announces the train to the ACC RBC, identifying

- the ETCS identity of the train,
- the ETCS mode of the train, and
- the identity of the border balise group where the train will enter the area of the neighbouring RBC.

The HOV RBC may send this information when the movement authority of the train ends at the border with the neighbouring RBC and the train sends a request for an extension of the movement authority, or when a route across the border is requested from the traffic control system.

If the ACC RBC can accept the train, it sends an acknowledgement to the HOV RBC. The ACC RBC may reject the train, on the grounds that it is not able to handle it currently. If this is the case, the ACC RBC will send a cancellation of the handover to the HOV RBC.

Step 2 (Route related information request)

In this step, the HOV RBC will send to the ACC RBC a route related information (RRI) request. This includes a first set of limits, if necessary, to ensure that the combined MA and other track description information sent by the HOV RBC to the ETCS on-board does not exceed the allowable limits specified in SUBSET-040. The RRI request will be repeated when the LRBG changes so that balise linking can be extended further into the ACC RBC area.

Step 3 (RBC-RBC transition is announced to the ETCS on-board)

In this step, the ETCS on-board is informed about the RBC-RBC handover. The information contains the communication information for the accepting RBC (namely the RBC identity and phone number) and the distance to the border location. This information is understood by the ETCS on-board as an order to contact the ACC RBC immediately if there is a free mobile radio available. In the case of only one mobile being available, due to a failed mobile, special conditions apply.

As this step triggers the communication between the ETCS on-board and the ACC RBC, the earliest at which it can be executed is when the train enters the geographical coverage of the radio network associated with the ACC RBC. Note that the ACC RBC may belong to a different radio network. This step may be commanded by the HOV RBC but can also be commanded by a balise group.

When the ETCS on-board is in direct communication with the ACC RBC, the ACC RBC may send information to the ETCS on-board directly. This is stored in a 'transition buffer' and not used before responsibility for the train has changed to the ACC RBC. For an ETCS on-board in Sleeping mode, there may be no need to contact the ACC RBC, and it is prevented from doing so.

The special case referred to above is when only one on-board mobile is available. According to the principles stated above, the on-board must not terminate the communication with the HOV RBC until the train has completely left the area of the HOV RBC. Only after doing that can the on-board mobile connect to the ACC RBC. Between terminating the communication with the HOV RBC, and while communication with the ACC RBC is being established, there is a gap of some seconds during which the train is not in communication with either RBC. The safety impact of this gap has to be analysed for each specific application. Consideration of the safety impact is also relevant in the scenario where the communication with the ACC RBC fails to establish successfully. Note that the HOV RBC is not aware whether one or two mobiles are available on-board. As a result, the decision when to connect to the ACC RBC is purely an on-board issue.

Step 4 (The ACC RBC becomes responsible)

The changeover of responsibility (i.e which RBC is responsible for supervising the train) is a decision of the ETCS on-board. The two circumstances in which it initiates this changeover are:

1) If two mobiles are available, when the train has passed the RBC/RBC border location with its 'maximum safe front end'.
2) If only one mobile is available, when the ETCS on-board has established communication with the ACC RBC.

From this moment on, the ETCS on-board will no longer accept movement authority or track description information from the HOV RBC, and any information received previously from the ACC RBC and stored in a transition buffer is used instead. When the ACC RBC next receives an updated position report from the train, it deduces that it has become responsible for the train because the train is now within its geographical area of control.

From now on the ACC RBC takes direct control of the train and informs the HOV RBC that it has taken over responsibility for doing so. In the case of two mobiles being available, communication with the HOV RBC is maintained by the on-board until the train has completely left its area. With the completion of this step, the on-board disconnects from the HOV RBC and the RBC/RBC handover is complete.

22.4 The communication protocol of the RBC-RBC interface

Communication between RBCs is in form of an on-going session, within which there may be several transactions, each dealing with one handover of a train between the two connected RBCs. At a given time, there may be trains passing between the two RBCs in different directions, and this may be at more than one border location.

The RBC-RBC communication protocol is structured into various layers, as shown in figure 22-2.

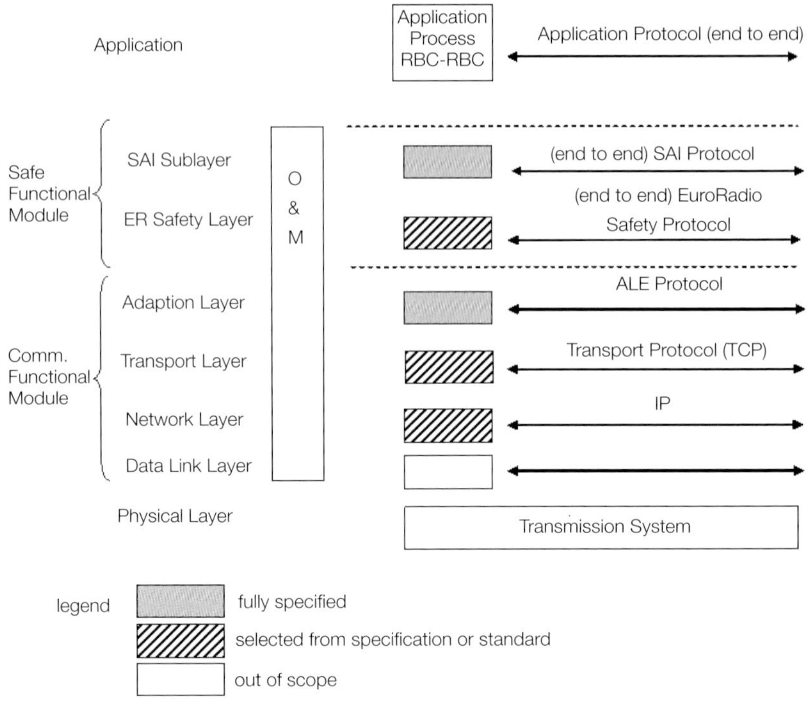

Figure 22-2: Architecture of the RBC-RBC communication system

The Physical Layer and the Data Link Layer are not harmonised (defined) by the RBC-RBC specification but will typically be Ethernet protocol using a standard transmission system.

The Communication Functional Module provides for:

- Redundancy to fulfil availability requirements
- Reliable, transparent and bi-directional transfer of data
- Retransmission of protocol data units, if necessary
- Monitoring of channel availability

Within this, the Transport Layer and Network Layer use the standard TCP/IP protocol. Above them is the Adaptation Layer, which manages redundant communication links and provides the interface with the Safe Functional Module. The Adaptation Layer provides two types of redundant links:

- Class A (optional): Two links are provided. Either of the two links can be used to transfer data, but the application layer packets are transmitted on only one of them (cold-standby).
- Class D (mandatory): Two links are provided, and normally all application layer packets transmitted on both links (hot-standby).

The Safe Functional Module has to detect and provide adequate defence against threats, as specified by EN 50159:2010. Within this, two safety layers provide the following common safety-related transmission functions:

- Message authenticity (origin and destination)
- Message sequence integrity
- Message timeliness
- Message integrity
- Reporting of safety relevant errors
- Configuration management (of the RBC-RBC safe communication protocol stack)
- Access protection

The first safety layer is EuroRadio, which is the same protocol as used for RBC to ETCS on-board communication via GSM-R, providing message authenticity (origin and destination), message integrity and access protection.

The Safe Application Intermediate (SAI) sub-layer provides the required additional protection against sequence error and delay threats by adding a sequence number field to the user data and either a Triple Time Stamp or Execution Cycle Counter.

References

SUBSET-039: FIS for the RBC/RBC Handover

SUBSET-098: RBC-RBC Safe Communication Interface

Chapter 23: RBC to Interlocking and other Control Centre Interfaces

23.1 Introduction

ETCS is not a complete system for railway control command and signalling, and a safety critical interlocking function is required to control points and lock routes through the network. This interlocking functionality is outside the scope of ETCS, as is the Traffic Control System, sometimes referred to as CTC (Centralised Traffic Control) or TMS (Traffic Management System), which provides the human interface with operators in the control centre and with the automation systems that manage train movements.

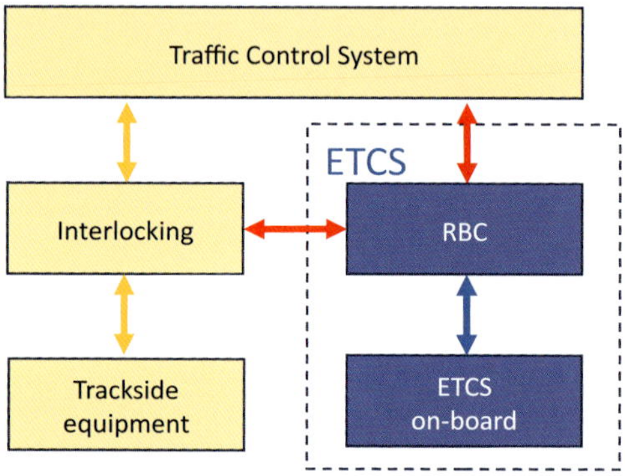

Figure 23-1: Interlocking and traffic control interfaces for an ETCS level 2 system

This chapter describes the interfaces between the RBC and these other control centre systems, as used in level 2 applications. Because the ETCS specifications focus only on the interoperability requirements and the traffic control system, interlockings and RBCs are all the responsibility of the infrastructure manager, there are no mandatory specifications for these interfaces.

The main objective of the communication from each interlocking to the RBC is to provide information needed by the RBC to calculate the movement authority for each train. In many cases the RBC and the interlocking are provided by the same supplier, and it is then possible for the two functions to be provided on a combined hardware platform.

23.2 RBC to interlocking – physical interface and communications protocol

The interfaces between RBCs and interlockings are usually based on IP networks that can be integrated with other IP networks used by the infrastructure manager. They must meet both the safety requirements (in accordance with the EN 50159:2010 standard) and the availability requirements for reliable service (usually involving duplication of the network and separation of the physical installation of those duplicated networks).

The specific interface that is used is dependent upon a number of factors, such as national constraints, project constraints, supplier constraints and the types of interfaces available in both the RBC and the interlocking subsystems. The last of these is of course a dominant consideration when interfacing with RBCs or interlockings that are already in commercial operation.

The choice of interface may also depend on whether both the RBC and the interlocking are provided by the same supplier or not. Where different suppliers are involved, three different methods have been used in practice. They are listed below as examples of possible solutions:

- Interfaces based on the SUBSET-098 interface that is specified for the RBC-RBC interface. This defines the communication protocol to be used, but not the nature of the application level of the interface.
- Use of the same type of interface as already exists to link interlockings with a national train control (class B) system and which provides similar services to those required by ETCS.
- Using a gateway as an intermediary between the interlockings and RBC, to provide the necessary physical connections and to convert the communication protocols on either side.

23.3 RBC to interlocking – functional interface

As with the protocol and physical interface, there is no mandatory functional specification for the interface between the RBC and the interlocking. This is because there is no harmonisation (standardisation) of the operating rules and principles for railways that use ETCS, the application of ETCS differs from country to country. This affects the detailed design and operation of the system, and one specific consequence is that the information exchanged between the interlockings and the RBC is dependent on the operating rules and principles applicable to each railway. For instance, information that a route is set and locked might be provided to the RBC in one of the following two ways:

- By giving the route status for each route that can be set by the interlocking. Each route status may be 'route not set', 'route set and locked – but track clearance not proven', 'route set, locked and accessible for main movements', etc.
- By giving the status for each infrastructure element controlled by the interlocking. These might be 'left/right/not locked' for points, 'occupied/clear/failure' for track circuits, 'open/closed/on-sight/shunting' and associated speed limits for lineside and virtual signals, etc.

While the main data flow is from interlocking to RBC, the passing of information from the RBC to interlockings is also necessary to derive benefit from the bi-directional radio communication between the train and the RBC. For instance:

- During the co-operative movement authority revocation process defined in the ETCS specifications, the RBC can provide the interlocking with an authorisation to release routes which have already been allocated to a train.
- The ETCS on-board can request a movement authority or an extension of a movement authority. This request is used by the RBC to initiate route setting in the interlocking when required by a train.
- Because the RBC knows the trains under its control, it can trigger a modification of the lineside signal aspect in the interlocking in certain circumstances – for instance to restrict entrance into a level 2 area to ETCS trains only; or to allow the use of multiple short intermediate block sections between lineside signals to improve performance.

23.4 RBC to traffic control system interface

The main channel for the exchange of information between the RBC and the traffic control system is via the interlocking. For instance, the operator or an automatic route setting system may request the interlocking to set a route, and the RBC will then generate the corresponding movement authority for the train. This means that an ETCS level 2 overlay on conventional lineside signalling can be implemented with minimal change to the signalling control workstation.

When ETCS level 2 is implemented without lineside signals, symbols representing the block section marker boards replace the symbols for signals on the workstation screens. If the interlocking receives status information from the RBC such as whether a movement authority has been accepted by a train, this may be indicated to the operator by another colour or an additional symbol next to the signal or block marker. The 'look and feel' of the workstation screens in the control centre will be specified either by the infrastructure manager or by the traffic control system supplier.

There is, however, status information in the RBC that is not related directly to the interlocking functionality and operation, such as train running numbers, train position, train speed and mode of the ETCS on-board. If any of this information needs to be made available to the operators and the automation systems in the control centre, a direct interface between the RBC and the traffic control system is required. This interface is also necessary if the traffic control system is to manage RBC functions such as temporary speed restrictions, or to send a text message or an emergency stop message to an individual train.

One way to satisfy this requirement has been for the ETCS trackside subsystem supplier to provide the operator in the control centre with a dedicated terminal connected directly to the RBC. It allows a control centre operator to view the information held in the RBC and enter commands. The interface is usually implemented on a personal computer. For some commands, such as removing a temporary speed restriction, a two-step process and associated procedural checks is required in order to achieve the required safety integrity level. In many cases this dedicated terminal is a variant of the facility normally provided for the maintainer.

This solution was acceptable for early ETCS implementations on dedicated sections of line, but for network-wide 'digital railway' implementation it is highly desirable to integrate this interface into the other systems in the infrastructure manager's traffic control centres. For example, integration of the operator interface allows the ETCS mode of each train to be highlighted by colour coding the border around each train running number on the workstation screens. Management of temporary speed restrictions can also be centralised, avoiding the risk of human error when re-keying information obtained from another system into a local RBC terminal.

Traffic control systems today usually include automated decision-making of the route setting process. In conventional automation systems this is limited by the granularity of train information available from the interlocking and the train detection systems. Significant improvements can be achieved by using much more detailed train position and speed data from an RBC to improve the prediction of future train conflicts and then optimise the route setting decisions to minimise such conflicts.

23.5 EULYNX

Although standardisation of the RBC-interlocking and RBC-traffic control system interfaces is outside the scope of the ETCS specifications, it is being addressed by the EULYNX initiative.

EULYNX is a European initiative of infrastructure managers to standardise certain interfaces and elements of the signalling systems. The project was started in 2014, with a planned three-year lifespan to complete the work. Subsequently the project evolved into a standing organisation for the standardisation of signalling system interfaces.

Because information technology components have a significantly shorter lifetime than robust field elements such as point drives and level crossings, the control centre electronics will be subject of more frequent replacements than longer-lasting signalling assets. EULYNX enables the disconnection of the life cycles of interlockings/control logic from those of field elements, facilitating a modular and more obsolescence-immune approach to the replacement of the control and automation system. This also makes market access easier for specialised manufacturers that may not supply the full range of control command and signalling assets, but which offer single subsystems such as level crossings and axle counters, and elements such as point machines. The objective of EULYNX is to establish long term stable interfaces and at the same time foster innovation at the module level.

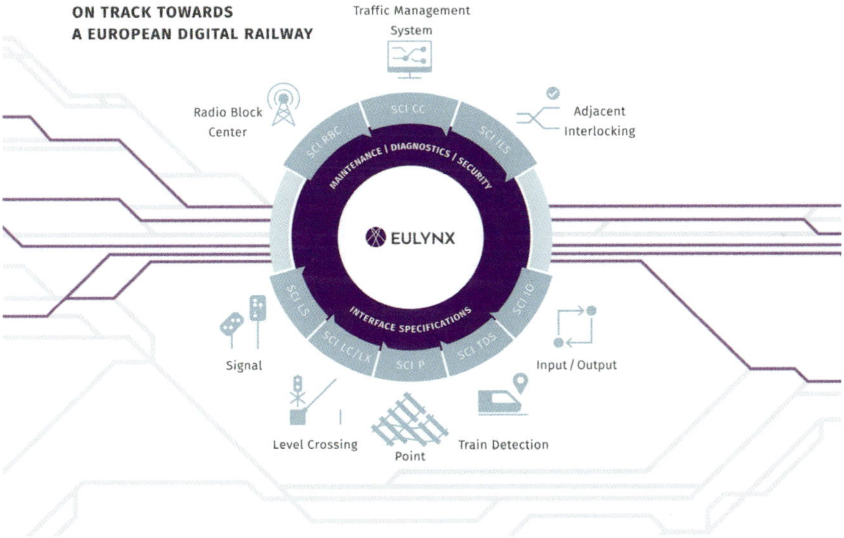

Figure 23-2: EULYNX system context

All of the EULYNX functional data interfaces (Standard Communication Interfaces – SCI) are based on a common process data interface based on IP networking, which includes a safety, retransmission and redundancy layer using the RaSTA (Rail Safe Transport Application) network protocol.

While a major focus of EULYNX is the interface between an interlocking and its associated field elements, thus reducing whole life cost by allowing interlocking renewal or upgrade while retaining existing field elements, the baseline specifications also include the interfaces between adjacent interlockings (SCI-ILS), interlocking to RBC (SCI-RBC) and interlocking to traffic control system (SCI-CC). These interfaces have been specified to be flexible enough to also cover other situations, including the following:

- SCI-RBC can be used for the interface from an interlocking to a centralised lineside equipment unit to control balises in ETCS level 1.
- SCI-CC can be used for the interface from an RBC to a traffic control system.

The EULYNX interfaces must be customised for different countries to suit the needs of the national operating rules and principles. For the interlocking to RBC interface, this customisation will remain necessary as long as there are different country-specific operating procedures, since these procedures are reflected in the application-specific design of interlockings. However, the EULYNX standard provides a common baseline onto which the country specific options can be built.

EULYNX also defines standardised interfaces for diagnostics, maintenance, and security, complementing the SCI used for functional data exchange. The diagnostic interface (SDI) enables event-driven and preventive data collection. The maintenance interface (SMI) supports engineering and configuration data management, allowing configuration and firmware updates, and also system resets. The security interface facilitates secure communication between systems, implementing identity and access management, public key infrastructure, and security logging to safeguard critical infrastructure. These interfaces, in addition to the SCI-type interfaces, ensure reliability, operational efficiency, and cybersecurity in railway signalling networks.

References

EULYNX documentation is available at https://eulynx.eu

RaSTA protocol is specified in German DIN standard VDE V 0831-200

Chapter 24: Certification and Testing of ETCS Constituents

24.1 Introduction

One of the fundamental concepts behind the standardisation of ETCS is that the interoperability constituents can be developed and 'placed onto the market' with an EC Declaration of Conformity that they comply with the requirements of the TSI CCS.

The Railway Interoperability Directive (EU) 2016/797 defines interoperability constituents (ICs) as meaning any elementary component, group of components, sub-assembly or complete assembly of equipment intended to be incorporated into a subsystem, upon which the interoperability of the rail system depends. Constituents may therefore be regarded as basic building blocks for railway subsystems. They include fixed equipment (infrastructure) as well as rolling stock equipment, and also intangible items such as software.

The ability to place ICs on the market with a Declaration of Conformity reduces the cost and timescales for ETCS applications by eliminating the need to re-validate the internal functionality and interfaces of the constituents for every project. The Declaration of Conformity and its supporting documentation of each constituent provides evidence that is used in the CCS on-board and trackside subsystem approval processes, and in the subsequent track to train integration and entry into service.

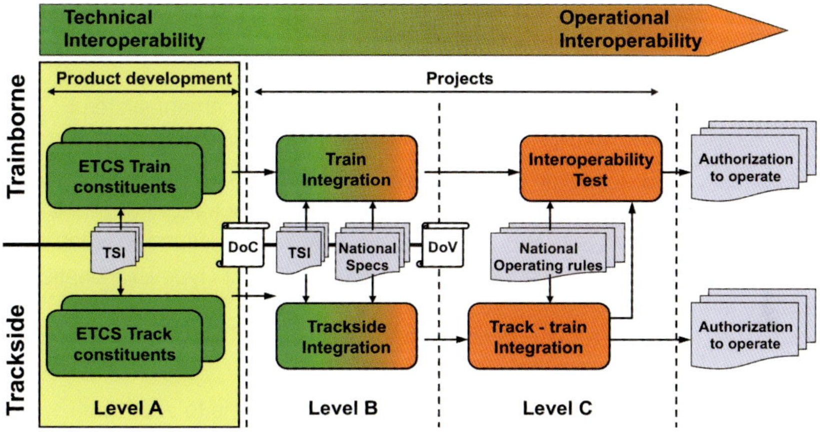

Figure 24-1: ETCS constituent certification within the overall approval process

The basic ETCS constituents are listed in the TSI CCS. They are:

- On-board: ETCS on-board, odometry, STM (ETCS interface only)
- Trackside: RBC, Eurobalise, LEU, Marker Board

The Declaration of Conformity process will also apply to other constituents covered in the TSI CCS but which are not part of the ETCS system, e.g. GSM-R, FRMCS, ATO.

The process for making a Declaration of Conformity is based on the European Union's common framework for marketing of products. It requires the manufacturer to employ an independent organisation known as a Notified Body to undertake a conformity assessment to verify that the constituent meets the requirements of the TSI CCS. Notified Bodies (NoBos) are certification agencies that are appointed by a member state of the EU, which notifies those appointments to the EU. In the context of ETCS their specific role is to verify the conformity of interoperability constituents and subsystems with the essential requirements of the TSI CCS.

In order to ensure that all Notified Bodies have a common understanding of the TSIs, ETCS specifications and assessment processes, they coordinate their work and consult with the ERA in a group called NB-Rail. The group's role includes resolution of any lack of clarity in the TSIs, implemented through so-called "recommendations for use" (RFU) – which are in fact mandatory despite the name.

This process is a legal requirement for ETCS applications only within the EU, but for projects elsewhere in the world the use of constituents with a Declaration of Conformity can still be a useful contribution to the approval process for ETCS projects.

24.2 Certification process

The activities a Notified Body will have to perform as part of its certification work depends on the so-called certification modules. These are defined in the European Commission decision 2010/713/EU, and the TSI CCS specifies which modules may be used and how they should be applied. The choice of the module(s) to be applied is subject to negotiation and agreement between the manufacturer and its Notified Body. The choice determines the type and validity of the certificate that can be issued. There are three options available, and they vary in terms of how far the NoBo is allowed to take account of the quality management system of the manufacturer. These are:

- Module CB (type examination) + CF (product verification)
- Module CB (type examination) + CD (production quality management system)
- Module CH1 (full quality management system with design examination)

In most cases either CB + CD or CH1 is chosen, as CF (product verification) requires the NoBo to inspect every example of the constituent that is manufactured. For all the options, the work is undertaken in two stages, with separate NoBo certificates issued for each of:

- the manufacturer's design and development of the constituent (Type Examination or Design Examination certificate)
- the production phase (Product Verification certificate or Quality Management System approval)

The purpose of the conformity assessment of a CCS constituent is to verify that:

- mandatory functions applicable to the interoperability of the constituent have been implemented
- optional functions applicable to the interoperability of the constituent have been implemented in line with the requirements
- any additional functions implemented are not in conflict with either the mandatory or optional functions applicable to the interoperability of the constituent
- any part of the constituent where the functionality is covered by national rules (e.g. open points in the TSI CCS, national functions in STMs, derogations) have been assessed and approved by the Member State
- construction of equipment (e.g. materials related to fire protection on rolling stock) and correct functioning within specified environmental conditions
- reliability, availability, maintainability and safety

It is worth noting that the certification process for interoperability by a Notified Body overlaps with the safety approval process. The TSI defines the tolerable hazard rate for random failures to be met for the trackside and on-board subsystems respectively, and also a set of specifications that are appropriate for compliance with the requirements of the Common Safety Method (CSM) for Risk Evaluation and Assessment (European Commission Regulation (EU) No 402/2013). The correct application of the risk management process is assessed by a CSM assessment body (AsBo), and the safety assessment report is used as evidence by the NoBo.

24.3 Certificate of conformity

In the case of successful verification, the Notified Body will issue EC certificates of conformity according to the modules applied for the conformity assessment, and also assemble a 'Technical File'. This contains a full definition of the constituent that has been assessed, the activities undertaken by the NoBo, and the evidence of conformity with the TSI. The Certificate of Conformity and the Technical File support the manufacturer or supplier's Declaration of Conformity that is required for the interoperability constituent to be 'placed on the market' and used in projects.

Any restrictions or conditions regarding the use of the constituent are recorded on the certificate of conformity. In some circumstances, a NoBo may issue an 'Intermediate Statement of Verification' to record what has been achieved at an intermediate project stage, and this can then be used as evidence to avoid duplication of effort by the same or another NoBo when completing the assessment at a later date.

Most constituent certificates remain valid as long as compliance of the constituent with the certification requirements is maintained. Ongoing validity depends upon showing that updated versions of the TSI CCS, error corrections etc, do not impact conformity with the basic parameters applicable to the constituent.

Certificates of conformity for Quality Management Systems have a validity of two years, and the NoBo determines how many surveillance audits are necessary within that interval.

If the constituent is still in production when its certificate of conformity expires, the certificate must be renewed. All changes to the original specification and design must be analysed, and it is the task of a Notified Body to judge whether the changes have an impact on the properties (e.g. interoperability, safety) on which the existing certificate is based. If so, the changes must be analysed, and conformity verified.

If a revision of the TSI comes into force, an interoperability constituent that already placed on the market remains valid for use in projects, unless the revised TSI explicitly states that this is not permitted.

24.4 Test laboratories

Practical testing in a laboratory is an important source of evidence to demonstrate compliance with the functionality and interface requirements for each interoperability constituent. For ETCS on-board interoperability constituents there is a specific requirement that tests are carried out in a laboratory accredited in accordance with EU Regulation (EC) No 765/2008, which means they are working to the ISO/IEC 17025:2017 standard for testing and calibration laboratories.

An accredited laboratory can be a testing department within an ETCS manufacturer, but there are also independent laboratories that provide this service. These have been important in the development of ETCS since the beginning, when they performed the first test campaigns with the manufacturers and suppliers in order to harmonise and consolidate the specifications for different critical components, such as the Eurobalise and ETCS on-board. They have also played a very important role in helping the UNISIG companies and the ERA to develop SUBSET-076, which prescribes the tests to be applied to ETCS on-board constituents for certification purposes. There are currently (2025) seven independent laboratories that are members of the ERTMS Accredited Labs association (EAL):

- CEDEX (Spain)
- DLR (Germany)
- ITALCERTIFER (Italy)
- LEF (France)
- Multitel (Belgium)
- RINA (Italy)
- Sofia Tech Park (Bulgaria)

For each constituent, the laboratory must provide a test bench that will simulate all the interfaces to the constituent under test, and a test management system that will execute the tests and record the results. Some of the interfaces are prescribed in detail in the ETCS specifications (FFFIS), whereas others are only specified at a functional level (FIS). In the latter case, the test bench must be adapted for the manufacturer's specific implementation of the interface.

The approach to defining the detailed test specifications varies depending upon the constituent that is being assessed for conformity. For the Eurobalise (SUBSET-085) and the ETCS on-board (SUBSET-076) there are detailed test specifications that are referenced directly from the TSI CCS. For other constituents (e.g. LEU and RBC) only the interface to other constituents is defined, and therefore detailed test specifications have to be developed which take into account supplier-specific functionality and interfaces.

24.5 Eurobalise testing

The requirements for Eurobalise testing are described in SUBSET-085. To ensure reliable operation of the air-gap in all conditions, a complex test set-up is required that allows the manipulation of the relative position of the balise under test in relation to the antenna representing the trainborne equipment and also tests the effects of solid and liquid debris on the train, metallic masses and cables in the vicinity. This requires a test rig that includes:

- antenna positioning tool (this could be a robot arm)
- reference loops
- test and activation antennas
- reference signal generators
- telegram generator
- reference receiver
- RF instruments and accessories
- reference units for solid/liquid debris, metallic masses, and cables

The tests performed on the test rig cover verification of the following:
- air-gap interface to the on-board antenna
- LEU interface for a controlled balise
- internal functionality
- crosstalk immunity with nearby cables
- compliance with specific EMC requirements

24.6 ETCS on-board testing

To demonstrate that an ETCS on-board is fully compliant with the SRS a large set of tests have to be performed. These are documented in SUBSET-076 and are defined in terms of features, test cases and test sequences.

A **feature** is a group of requirements, compliance with which can be tested at the available interfaces. Features are the characteristic behaviour expected by the operator from a system, and are therefore outwardly visible. Examples of test features are 'Balise information types and usage' or 'Structure of a movement authority'. For Baseline 3 release 2 (SRS 3.6.0) there were 193 features.

For each feature, various **test cases** are defined. For each test case the following requirements are defined:
- target to be achieved
- methods of measuring results
- constraints
- starting conditions (internal to the ETCS on-board and at the interfaces)
- steps to conduct the test

For each step, the interface event that triggers the step and the expected outcome of the step are described, together with any change of ETCS level/mode. Each feature is tested in several test cases, and each test case provides traceability of the SRS requirements that are assessed. This traceability between the test cases and the SRS requirements provides assurance that application of all the test cases will verify the fulfilment of all the ETCS on-board requirements specified in the SRS. For Baseline 3 release 2 (SRS 3.6.0) there were 1800 test cases.

A **test sequence** is used to enable automatic execution of the test cases by combining them into scenarios that can be run within the ETCS on-board reference test facility environment. Each test sequence is defined as a complete simulated train trip, starting and finishing with the train stationary and the ETCS on-board in No Power (NP) mode. Both the trip speed profile and the messages (radio, loop and balise) sent to the ETCS on-board are completely defined in each test sequence. The complete set of test sequences covers all test cases and, therefore, all SRS requirements. For Baseline 3 release 2 (SRS 3.6.0) there were 775 test sequences.

The **test environment** for the ETCS on-board is specified in SUBSET-094 (Functional Requirements for an on-board Reference Test Facility). The test environment provides all the interfaces to the ETCS on-board, to enable simulated operation of a train on which the ETCS on-board has been installed. For the interfaces that are not defined at FFFIS level in the TSI CCS, an interface adaptor is needed to connect the supplier-specific implementation to the reference test facility. The interface with the DMI uses image recognition software to record the information displayed on the screen, and it simulates human input such as data entry and acknowledgement of warnings using a 'robot finger' interacting with the touch screen.

Test sequences are loaded into the system using the Laboratory Scenario Editor (LSE). To execute a test sequence, the required telegrams, messages and other inputs to the ETCS on-board required at each step are passed to the Laboratory Scenarios Controller (LSC), and corresponding requirements for train movement to the Train Movement Simulator (TMS). The TMS calculates the real-time movement of the train based on a model of the train dynamics and the infrastructure and provides a continuous update of speed information via the Speed Sensor Simulator (SSS). At the trigger points specified in the test case (which are usually locations on the simulated track), the LSC will send the specified balise telegram, radio message or other input update to the ETCS on-board via the appropriate interface adaptor. All of the information passing to and from the ETCS on-board is recorded on the Laboratory Event Recorder (LER). This includes the information saved on the ETCS on-board's own juridical recorder (JRU). This information is time stamped and correlated with the simulated train position.

The analysis and evaluation of the executed test sequence then compares the DMI, JRU and TIU outputs with the expected values. A first analysis of the result of the test is performed with the Automatic Evaluation Tool (AET). The output from this tool is reviewed by expert personnel to determine whether the test has been completed successfully, or whether any deviations or non-compliances have been observed. The results of the full suite of test sequences are then complied into a report which forms the basis for certification of the ETCS on-board by a notified body.

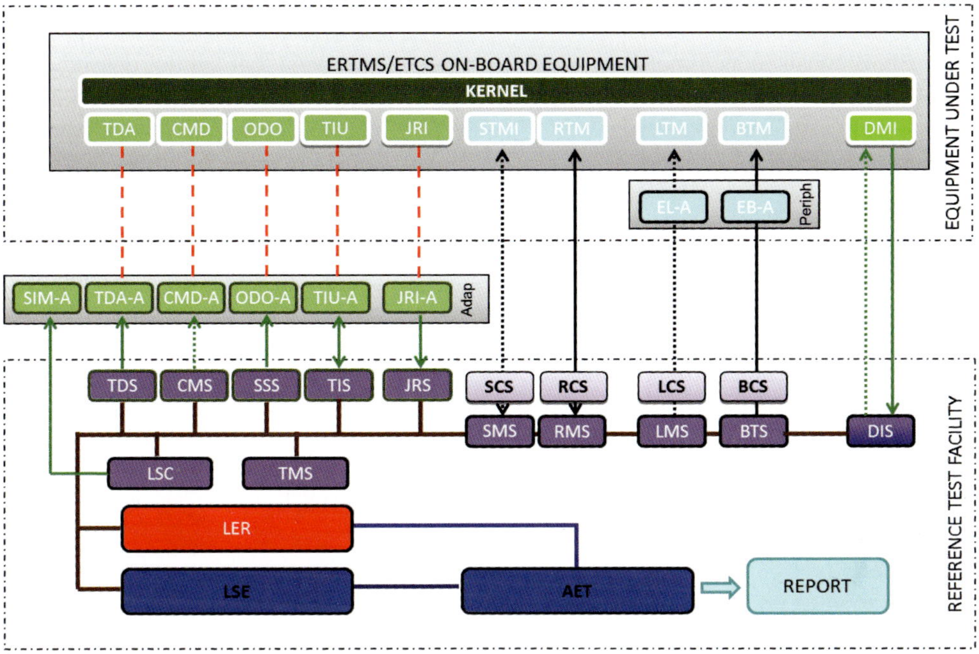

On-board equipment	Interface adaptors	Interface simulators	Test management
RTM – Radio Transmission Module	RCS – EuroRadio Communication Simulator	RMS – EuroRadio Messages Simulator	SIM-A – Simulation Management Adaptor
BTM+EB-A – Balise Transmission Module & Antenna	BCS – Eurobalise Communication Simulator	BTS - Eurobalise Telegram Simulator	LSC – Laboratory Scenario Controller
LTM+EL-A – Loop Transmission Module & Antenna	LCS – Euroloop Communication Simulator	LMS – Euroloop Messages Simulator	TMS – Train Motion Simulator
STM – Special Transmission Module	SCS – STM Communication Simulator	SMS - STM Messages Simulator	LER – Laboratory Event Recorder
JRI - Juridical Recording Interface	JRI-A – JRI Adaptor	JRS - Juridical Recording Simulator	LSE – Laboratory Scenario Editor
TIU – Train Interface Unit	TIU-A - TIU Adaptor	TIS – Train Interface Simulator	AET – Automatic Evaluation Tool
ODO – Odometry Function	ODO-A - ODO Adaptor	SSS – Speed Sensor Simulator	
CMD – Cold Movement Detector	CMD-A – CMD Adaptor	CMS – Cold Movement Simulator	
TDA – Train Data Acquisition	TDA-A – TDA Adaptor	TDS – Train Data Simulator	
DMI – Driver Machine Interface		DIS – DMI Interface Simulator	

Figure 24-2: Reference test architecture for ETCS on-board equipment

24.7 Testing of other ETCS constituents

Other ETCS constituents (e.g. RBC, LEU) have external interfaces that are not defined in the TSI CCS, and therefore the test requirements cannot be fully described in the specifications. It is the responsibility of the manufacturer or supplier, the notified body and the test laboratory to agree on a test environment and a test process that will demonstrate compliance with the relevant requirements of the ETCS System Requirements Specification (SRS) and any other applicable standards.

For a radio block centre (RBC) the test environment will include software simulation of one or more trains (including both ETCS on-board and train dynamics), the interlocking, the trackside equipment and the traffic control system for a simulated area of railway that has all the features necessary to exercise and test the required functionality. Unlike the ETCS on-board which is required to implement the full functionality specified in the SRS as a generic product, testing of an RBC is targeted at a generic application, i.e. to work with a specific type of interlocking and with a specific infrastructure manager's signalling principles and operating rules. The process of testing using pre-defined test sequences, recording of the responses and analysis and reporting of the results is similar to that for the ETCS on-board.

References

ISO/IEC 17025:2017: General requirements for the competence of testing and calibration laboratories

SUBSET-076: ETCS On-board Test Specifications

SUBSET-085: Test Specification for Eurobalise FFFIS

SUBSET-094: Functional Requirements for an on-board Reference Test Facility

Website for the NB-Rail Association: https://nb-rail.eu

Website for the ERTMS Accredited Labs organisation: https://ertmsaccreditedlabs.org/

Part 4
ETCS Deployment in Projects

Part 4 describes the issues that arise when deploying ETCS on a specific railway line or a fleet of trains. In this book we have not attempted to describe any specific projects, so the process is described in generic terms, from strategic planning through detailed design of the on-board and trackside applications to installation, testing, certification and entry into service. The important issues of operational readiness, training of drivers and other staff to work with the system and in-service support are covered.

Chapter 25: Planning for ETCS Deployment

25.1 Operational rules and engineering standards

The ETCS standards allow suppliers to develop ETCS products (systems, sub-systems and constituent parts) that are applicable to a wide range of applications and that will be interoperable with other suppliers' products. However, the application of ETCS products to specific projects is not so straightforward, for two main reasons:

1. The ETCS standards focus on the interoperability between the trackside and the on-board and the on-board interfaces. Key elements of trackside signalling such as interlockings, traffic control and interfaces to equipment on the track are excluded and thus are often unique to a particular supplier.
2. Unless a completely new self-contained railway is being constructed, there will be existing operational rules and engineering standards that need to be reviewed and adapted where necessary.

With its various levels and modes of operation, ETCS is a highly adaptable 'toolkit' that can be applied in many different ways. This is a key strength that allows it to be applied to all types of railways and to replace a variety of existing train protection systems. But this strength also creates a risk that every project devises a slightly different and unique solution, even within a single railway administration. For this reason, it is important that each infrastructure manager (IM) puts in place the necessary operational rules and engineering standards to ensure a consistent application across all ETCS projects within the IM's area of responsibility. This must be a collaborative approach that also involves the railway undertakings (RU) and national safety authority (NSA) as well as the suppliers of the ETCS and other systems.

In most cases the starting point will be a national strategy for ETCS deployment, which for EU member states will have been defined in a national implementation plan. In some cases, the intention may be to retain the existing trackside signalling and simply replace an existing national train protection system with ETCS. In others there may be more ambitious plans to migrate to cab signalling with minimal use of trackside equipment. In all cases it is important to have a pragmatic migration strategy that focusses on achieving a clear end goal with, of course, the necessary funding.

One approach to developing and validating the operational rules and engineering standards is through the development and definition of operational scenarios, using an 'operational simulator' to visualise their practical application. This is typically a desktop computer running software similar to a driver training system. It comprises the ETCS on-board and DMI functionality together with a view of the track ahead and a means of scripting the messages between ETCS trackside and on-board as the simulated train moves along a pre-defined track. This type of environment is extremely useful to give all stakeholders a clear understanding of what is being proposed in the new rules and standards, and to ensure that any problems are identified and addressed before the project enters its design and implementation phases.

The operational scenarios should cover, as a minimum:

- Normal train operations over a variety of track layouts
- Operational interactions between trains including one train following another, splitting trains, joining trains, and shunting
- Deviations from normal working, e.g. temporary speed restrictions, emergency stop
- Failure modes of both ETCS and other railway systems, and failure management

- Transition between ETCS levels, including to and from national train control systems
- Intermediate states of the system that will occur during the migration to ETCS and which may require specific operational rules and methods

For most railways it is inevitable that the operational rules for ETCS train working will be adapted from the existing national rules. This has meant that in practice ETCS is being used in different ways in different countries, and that barriers to full interoperability exist at some international borders, even when ETCS is in use on both sides of a border.

There have been several initiatives to reduce the country variations, including the definition of standard 'written orders' to be used by train drivers and traffic controllers when giving and receiving verbal instructions to make train movements in degraded conditions. Where ETCS is applied as an overlay to existing lineside signalling, some variation between countries is inevitable, but for lines with cab signalling the standardisation of operational rules may eventually be possible across Europe.

Alongside the development of the operational concepts and rules, the infrastructure manager will also have to amend or supplement the existing signalling principles and standards to take account of ETCS deployment and to define the basis on which suppliers will contract to deliver specific projects. The issues that will need to be addressed include:

- Definition of ETCS national values to be used
- Application of ETCS modes to interlocking route types (classes) and signal aspects
- ETCS interactions with level crossings
- Adapting overlap concepts and rules to work with the ETCS concepts of supervised location and release speeds at the end of a movement authority
- Use of ETCS features such as emergency stop and the co-operative shortening of a movement authority, including how these features affect interlocking functionality
- Use of ETCS track condition messages not directly linked to signalling, e.g. traction changeovers, special brakes that must not be used in certain locations, information about tunnels and stations
- Policy for provision of ETCS marker boards and other lineside signs
- Additional ETCS information required on signalling scheme plans and other documentation, e.g. showing location of balises and end of authorities
- Provision of a key management centre and a policy for updating keys and their secure offline/online delivery to both onboard and trackside sub-systems
- Arrangements for transitions between ETCS and non-ETCS lines

Similar considerations apply to the railway undertakings, who will need to consider the application of ETCS on the types of train and services that they operate:

- Which special modes (Passive Shunting (PS), Non Leading (NL), etc.) are relevant to their operations
- How to incorporate ETCS procedures such as start of mission into existing driver tasks
- Managing values such as train length and braking curves, especially for variable formation trains

The ERTMS Users Group has provided a forum to collect and analyse these issues in order to learn and share experiences about ETCS application design, and they have published a series of Engineering Guidelines providing harmonised implementation recommendations on various topics.

25.2 System compatibility checks

When the details have been defined for the application of ETCS in a specific area (or route), the infrastructure manager must define the ETCS System Compatibility (ESC) and Radio System Compatibility (RSC) checks to be conducted to demonstrate that an ETCS-fitted train will operate reliably with the trackside ETCS and RMR in the area. These may be document checks, laboratory tests or track tests to be performed, and are additional to the requirements specified in the TSI CCS. The area of use to which a specific set of checks applies is known as an 'ESC/RSC Type'. This might be a specified geographic area or a category of line within the infrastructure manager's network.

Within the EU each infrastructure manager submits their ESC/RSC check requirements to the ERA, which publishes a list of all ESC/RSC types together with references to the documents where the associated checks have been defined. This list and the referenced documents are then used by a railway undertaking to determine what they need to do to obtain approval for their ETCS-fitted trains to operate over any specific geographical area or route. Where the ESC/RSC checks include specific tests, the infrastructure manager must provide the laboratory or track access to enable these to be undertaken.

25.3 Migration planning

A crucial element of strategic planning for ETCS is the phasing of ETCS fitment on trains and trackside, and decisions on whether dual fitment of ETCS and national train control systems is required as part of the migration.

Dual fitment of the trackside is possible if lineside signals and train detection are retained (an ETCS overlay solution), but this means that the potential cost savings and full performance benefits of ETCS cannot be realised. Avoiding dual fitment of the trackside on any specific route avoids these problems, but it requires that every train that needs to run on the route is equipped with ETCS on-board (and the drivers trained) before the ETCS trackside is commissioned. If the long-term goal is to operate at level 2 without lineside signals, an overlay solution could be useful at the start of the programme to provide a portion of the route for ETCS testing and driver training while trains continue to operate using the national train control systems, but this should be avoided as a widespread solution. If a longer-term overlay solution is required to accommodate unfitted trains that relatively infrequently use the route, it may be possible to reduce the number of lineside signals and accept a capacity limitation when an unfitted train is operating.

The optimum solution may be to programme the ETCS trackside installation to follow the delivery of new ETCS-fitted trains on the route. This may be practicable for a dedicated suburban, high speed or freight only line, but it cannot be applied to parts of the national network over which multiple railway undertakings operate a variety of train types. It may also cause issues for the maintenance and update of the ETCS on-board if there is an extended period where it is not in use while awaiting the fitment and commissioning of the ETCS trackside.

In most cases the trains, whether new or existing, will be dual fitted using an STM or another approach as described in chapter 20. For new trains, fitment is relatively straightforward. It is now mandatory within the EU for all new trains to be fitted with ETCS, and most train builders have an in-house supplier of ETCS on-board equipment that they use.

Retrofitting ETCS to existing trains is a more complex engineering challenge and an expensive process, due to the impact on the existing safety-critical on-board systems and the requirement to take trains out of service for installation and testing of each train. This is especially true for small volume bespoke fitments where there is no economy of scale.

The typical lifespan of a train is 40 years and therefore retrofitting of ETCS to existing trains is often unavoidable. Freight locomotives and track maintenance vehicles usually operate widely across a rail network so it is inevitable that some of these will need to be retrofitted at an early stage in an ETCS implementation programme. There may also be a requirement to fit heritage locomotives that operate occasional charter trains.

On-board ETCS fitment is not only an engineering challenge but also a commercial one, particularly where the infrastructure manager and the railway undertakings are independent entities, for example, as required by EU legislation. The railway undertakings incur additional up-front costs to install ETCS on-board on their trains, whilst trackside equipment cost savings and performance enhancements generally benefit the infrastructure manager. This means that a commercial or legal incentive needs to exist for the railway undertaking to fit ETCS. Within the EU, this is contained in the TSI CCS, which places a legal requirement on the railway undertakings to install ETCS on new trains and gives the infrastructure manager the right to restrict access to ETCS-only trains on designated routes, provided that 5 years' notice has been given to the affected railway undertakings. Outside the EU, there may need to be equivalent national legislation or a commercial arrangement such as reduced track access charges for ETCS-fitted trains to provide an incentive for fitment. Of course, the situation is easier where there is a single vertically integrated railway operation.

References

ERTMS Users Group: Engineering Guideline
Available at https://ertms.be/activities/engineering-support

European Union Agency for Railways: ESC/RSC Technical document
Available at https://www.era.europa.eu/content/etcs-and-radio-system-compatibility-escrsc

Chapter 26: ETCS Trackside Engineering

26.1 Infrastructure data

One of the major goals of ETCS is to allow trains to operate over different lines within a country or several countries without the need for the ETCS on-board to be pre-programmed with all the data about all the routes on which those trains are to be operated. The permissible speeds, track layout and interaction with the signalling elements are included within the configuration data of the trackside elements (LEUs, RBCs, balises) and used to generate the movement authorities that provide all the information required to supervise the safe movement of the trains.

This means that the track data preparation process is critical from a safety point of view, as the behaviour of the ETCS on-board depends upon the accuracy of the track data received from the trackside. To achieve the required level of safety integrity, the data preparation process and the accuracy of the source information on which it is based must be commensurate with the SIL 4 level of safety integrity, which is the level required for the most critical functions of ETCS.

The starting point for the process must be a very accurate survey of the railway track layout and the location of relevant lineside equipment. Once obtained, this survey data must then be maintained and updated to take account of any infrastructure changes that are happening during the ETCS deployment and, of course, during the subsequent operational use of ETCS.

Modern survey techniques such as LIDAR (light detection and ranging) point cloud scanning using drones or train-mounted equipment can now achieve millimetre positioning accuracy (a point cloud is a collection of points of data plotted in 3D space). From the ETCS viewpoint the requirement is a topological map that defines the railway in terms of nodes and links, with a precise length and a gradient profile for each link. Locations such as the start and end of speed restrictions, as well as the locations of physical entities such as switch toes (points), can then be defined as a position along a link, and where required the direction of train movement to which they apply.

Various specifications and standards have been developed that are useful for the exchange of this type of information:

- EULYNX is an initiative to standardise interlocking functionality (see chapter 23) and interfaces. It includes a data exchange format.
- railML is an XML-based data exchange format for data interoperability of railway applications.
- SUBSET-112 is part of the ETCS Interoperability Test Scenarios Specifications, although it is not published as a formal standard by the ERA. It defines a standard way of describing the track.

Railway projects increasingly make use of Building Information Modelling (BIM) standards to co-ordinate activities between all the disciplines in major construction projects and subsequent asset management. BIM is based on a detailed 3D (three-dimensional) model of the whole railway, including track, structures, earthworks, electrification equipment, buildings and cable routes. In signalling projects BIM has been used to determine the best location of signals to ensure good visibility, and to generate cab view graphics for driver training simulators.

It should be noted that BIM provides a 3D vision of the track and other elements, and this needs to be transformed into the topology, distance and gradient information required by ETCS. Despite this, there are significant benefits in ensuring that the ETCS design is aligned with other engineering changes on the railway.

26.2 Balise location

One of the key tasks at an early stage in the application engineering for an ETCS project is to determine the location and functionality of balises to be installed on the track. As balises provide the reference points for all train positioning information in ETCS, their location within the track topology is a key input to the data preparation process. Balise groups are provided to send messages to trains at specific locations and to provide a location reference for the ETCS on-board.

For ETCS level 1, the basic requirement is that a balise group is located at each signal that can display a stop aspect. In most cases this will be supplemented by additional infill balise groups on the approach, as described in chapter 7. One balise in each group will be a controlled balise connected to a LEU, which provides the movement authority information consistent with the aspect displayed by the signal.

Other balise groups in level 1, and most balise groups in level 2 will be fixed balises that provide a position reference or other static information. These will include balise groups for purposes such as:

- Limits to movements in Shunting (SH) or Staff Responsible (SR) modes.
- Level transitions, e.g. on approach to a level 2 area there may be balises to trigger, network registration, radio connection, transition annunciation, transition border as described in chapter 9.
- To allow the ETCS on-board and RBC to establish an unambiguous position for the train in any train awakening area. This is any area where an ETCS-fitted train will be expected to power up its ETCS on-board and start up the registration process. These might be depots, sidings or stabling points.

In level 2 the majority of balise groups are provided for train positioning purposes only. Their location and spacing will be determined by how accurate the train positioning needs to be. Closer spacing will be required in areas where a wide margin between the actual train position and the on-board calculated position (minimum/maximum safe position of the front and rear of the train) due to cumulative odometry error cannot be tolerated. An example of this would be where a train needs to stop very close to an end of movement authority to ensure the rear of the train is alongside a station platform or is clear of another route. Another example is where ATO over ETCS is deployed, in which case additional balises may be needed to ensure station stopping accuracy.

Reliable data transmission between the ETCS on-board and a balise may be degraded by the presence of large metal masses in the vicinity of a balise. Such masses could include bridge structures under the track, steel sleepers, guard rails and electric traction impedance bonds or cables. Detailed application rules for balise locations are specified in section 5.7.10 of SUBSET-036 "FFFIS for Eurobalise", and if these cannot be respected then the balise location will need to be moved away from the large metal mass, which may then require other compensatory changes. There are also rules for the spacing between balises and the routing of cables in their vicinity.

26.3 Signalling plans for ETCS

As for any conventional signalling scheme, a key document to be generated at an early stage of an ETCS deployment project will be a signalling plan that shows a schematic layout of the railway track in the area, and the location and naming of all the features that are relevant to the signalling. This will include physical objects such as balises, points, signals, signs and train detection section boundaries. For ETCS purposes it may also include other locations that are linked to specific functions, such as mode/level transitions.

As the design progresses, more detailed signalling specification information will be generated, often in a tabular format. An important element of this will be the definitions of the interfaces with the elements of the control command and signalling system outside the scope of ETCS, for instance the interlocking and traffic control system. The format for that will depend on the ETCS level. Level 1 will require the design of an electrical interface to link LEUs to lineside circuits or a data link from each LEU to the relevant interlocking, whereas in level 2 there will be a direct data interface between RBC and interlocking, which are often co-located in the same building.

26.4 Data preparation processes and tools

Every supplier of ETCS systems will have their own set of computer-based data preparation tools to produce the detailed data configuration for balises, RBCs and other ETCS trackside system components.

The data has to be generated with a level of accuracy commensurate with the SIL 4 safety integrity required for the most critical functions of ETCS, and this needs to take into account the possibility of human error in entering data into the tool, as well as errors in processing the data by the tool, and at each stage of the process up to and including the loading of the data into the ETCS hardware.

Achieving this level of integrity cannot simply rely on the competence of the data preparers, and the integrity of the data processing software – the overall process needs to incorporate rigorous verification, validation and configuration management to provide sufficient confidence in the output data. This may include an element of traditional document review and principles testing by experts who are independent of the design team, although automated tools that can verify the consistency of the output data with the source data (requirements), and against safety principles and design rules, are also important.

References

SUBSET-036: FFFIS for Eurobalise

SUBSET-112: UNISIG Basics for Interoperability Test Scenario Specifications

Chapter 27: ETCS On-board Train Installation

27.1 System architecture and constraints

The basic system architecture of the ETCS on-board and its interfaces is described in chapter 20 and applies to all trains, but there are differences in the detail of the arrangements, depending on the type of train. There are many variants, but some typical examples are:

Train type	Locomotive hauling freight wagons or passenger carriages	Short multiple unit train (driveable from either end)	Long fixed formation train (driveable from either end)
Number of on-board units	One	One	Two (one in each end vehicle)
Number of DMIs	Two (one in each driving cab)	Two (one in each end vehicle)	Two (one in each end vehicle)
Number of balise antennas	One balise antenna in centre of vehicle	Two balise antennas (one on each end vehicle)	Two balise antennas (one on each end vehicle)
Data entry	Train data entered manually by the driver	Train data provided automatically by train systems, depending on how many units are coupled together	Train data fixed in ETCS on-board

Table 27-1: Examples of ETCS on-board architecture

Various factors and constraints must be taken into account during the engineering process, including:

- Environmental conditions – temperature and humidity, vibration and shock conditions, fire protection, dust and water ingress
- Electromagnetic compatibility – applicable standards, requirements for shielding and earthing contacts, earthing of the hardware, use of interference filters
- Cabling between the components – types of cables, isolation and dielectric strength, shielding, routing within the vehicle, inter-vehicle connections, terminals
- Power supply – continuous and transient currents drawn, rating of circuit breakers, cable dimensions, voltage convertors, filters, independent supply for cold movement detector
- Physical space – available locations for ETCS equipment, relocating existing equipment where necessary, equipment fixings and enclosures
- Existing train systems – train management system or hard-wired controls, type of brake interface (electrical or pneumatic), sharing of interfaces with other functions, e.g. Class B national train control or driver vigilance systems
- Human factors – cab layout, ergonomic considerations for the driver, accessibility for maintenance

27.2 ETCS on-board cabinet

The ETCS on-board hardware usually comprises a number of modules that are mounted in a rack. This needs to be installed in a cabinet that provides the necessary mechanical, environmental and EMC protection for the equipment. The location of the cabinet needs to be somewhere that will be accessible for maintenance but also protected against unauthorised access. If a STM is provided to interface with a Class B national train control system, this will usually be installed in the same cabinet as the ETCS on-board.

As an alternative to using a single cabinet for the equipment, some suppliers have designed their on-board equipment as a number of discrete modules that can be distributed in smaller spaces around the vehicle.

The ETCS on-board is linked by data cables to the other ETCS components on the train and also needs to be interfaced to non-ETCS elements such as the brake system. As described in chapter 20, this can either be via a standardised data interface or via individual hard-wired relay contacts for each function. The most commonly used functions (M = mandatory) are:

- Inputs to the ETCS – isolation (M), cab status (M), direction controller, sleeping
- Outputs from the ETCS – emergency brake (M), service brake

27.3 Cab equipment

The main element to be installed in each driving cab of the train is the DMI. The location of this is critical as it needs to be directly in front of the driver so that the displayed information is easy to read at all times without distracting from the view out of the cab window. It also needs to be within easy reach of the driver for data entry and push button acknowledgements. This requires a human factors analysis to inform the design, to take into account variations in eye height and arm length of the driver and the adjustability of the driving seat. The angle, intensity and type of lighting is also critical, to avoid reflections on the DMI screen in all weather conditions and times of day.

These requirements mean that the DMI is normally installed in place of a conventional speedometer directly in front of the driver. This implies that in the event of DMI or ETCS on-board failure, the driver has no information about the speed of the train. For this reason, it is usual to provide a backup facility, either in the form of a conventional speedometer or a duplicate DMI elsewhere in the cab. One option is for the duplicate DMI to be displayed on a screen provided in the cab for another purpose, such as the train management system. Another option is to provide a 'split screen' ETCS DMI which normally displays the main speed display and the planning area on separate smaller screens, with the ability to move the speed display to the second screen in place of the planning area if the first screen has failed (see figure 29-1 for an example).

In normal operation the DMI and the existing cab controls (e.g. cab status, direction) provide the complete interface with the train driver. The only additional facility required is the ability to isolate the ETCS system in the event of a failure that is preventing the movement of the train. This needs to be somewhere accessible to the driver, and is usually located in the cab, very often in a sealed location to discourage its use unless absolutely necessary. Once ETCS has been isolated, an alternative means of providing a speed indication to the driver will be required, since the ETCS cannot provide the information. This could be a separate feed to the DMI independently of the ETCS on-board, or a backup speedometer via another train system.

27.4 Balise antenna

The Eurobalise antenna needs to be mounted under the train. The location needs to be carefully chosen in accordance with the installation rules set out in SUBSET-036 and SUBSET-040. These rules address the following requirements for the location:

- Centrally located between the running rails
- Minimum and maximum height above the rails, taking account of dynamic movement, wheel wear, etc.
- Minimum distance from front of the train (2 m to ensure the separation between antennas of two trains coupled together)
- Maximum distance from the first axle of the train (12.5 m for compatibility with the rules for positioning level 1 balises at a train detection boundary)
- Metal-free space required around the antenna
- Away from any other antenna or high current electrical equipment
- Protected against flying gravel, high pressure water thrown up off the track and from rain and ice
- The routing of the cables between the antenna and the ETCS on-board

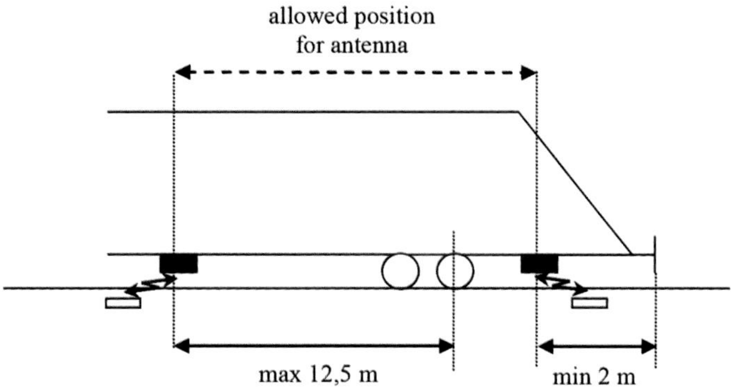

Figure 27-1: Location of balise antenna on the vehicle (from SUBSET-040)

For a single vehicle with two cabs, the distancing rules allow for a single antenna to be mounted in the centre of the vehicle.

If the vehicle is fitted with a national train control system (Class B), then that will require its own antenna or other device to communicate with its trackside equipment, and an appropriate spacing between this and the Eurobalise antenna will be required to ensure both systems will operate reliably.

27.5 Speed and distance measurement equipment

Most ETCS systems use an axle-mounted odometer that senses the rotation of the wheels as the primary method of speed and distance measurement. For availability, and to provide a degree of mitigation against errors due to wheel slip in low adhesion conditions, two or more axles may be fitted with odometers, both connected to the ETCS on-board. The choice of which axles to fit should take into account the various risk factors that lead to wheelslip. The ideal arrangement is to use an axle that is not used for either traction or braking in normal operation, but in most cases an unbraked axle is not an option. It is also best to avoid an axle at the front of the train, as this encounters the lowest adhesion conditions. The axle end environment, where the odometer is mounted, is a particularly harsh one in terms of exposure to shock, vibration, dust and moisture, so high quality connections and careful routing of cables are essential to ensure reliability.

To mitigate against wheelslip, most ETCS applications use sensor data fusion, which combines the odometer data with that from another system which is not affected by wheel-rail adhesion. Technologies used for this include doppler radar, optical spatial filtering velocimetry and inertial measurement using accelerometers. Each of these will require appropriate installation on the vehicle. For instance, a doppler radar will need to be mounted somewhere under the train so that the radar beam can be directed towards the track. The location needs to be selected so as to avoid any moving parts of the train that could produce false reflections of the radar beam, e.g. wheels, brake discs, drive shafts, flexible cables and pipes, and any train-generated or externally sourced electromagnetic interference. As with other underframe mounted equipment, robust installation arrangements are required to withstand the harsh conditions.

27.6 Radio antenna

Each GSM-R data radio used for ETCS purposes will need an antenna on the roof of the train. The ETCS on-board typically uses two GSM-R radios, and there will also be a GSM-R voice radio. Their antennas all need to be spaced apart to avoid interfering with one another. They may also need to be appropriately spaced from other roof-mounted antennas (for instance those used for public mobile radio systems or GPS/GNSS satellite positioning) and from the high voltage pantograph on an electric train. In future there will also need to be provision for additional or alternative antennas for FRMCS.

27.7 Data configuration

In addition to the physical installation on the vehicle, it is necessary to configure the supplier's standard ETCS on-board product for the specific application. The information required to do this usually takes the form of data in tabular format, or an application-specific language which is interpreted by the generic software.

Most of this data is fixed for a specific type of vehicle and is validated in a 'first of class' installation, verification and testing process before being replicated in every vehicle of the same type. Examples of the types of data required to configure the ETCS on-board include:

- For configuration of the train interface, functions used, data interface or hard-wired
- For configuration of the ETCS constituents, DMIs, antennas, STM, etc.
- To set limiting values of the train data, including data that is fixed and data that can be provided by driver data entry
- Service brake, emergency brake and traction cut-off characteristics

There will also be a few parameters that need to be configured individually for a specific vehicle and possibly updated during the vehicle life. A typical example is wheel diameter, which is critical for accurate speed and position measurement. Wheels wear down in service and so the diameter reduces gradually. Some suppliers provide an automatic calibration function that accounts for this, but a manual update will still be needed when wheels are replaced or re-profiled on a wheel lathe.

As has already been mentioned with reference to the trackside equipment, the ETCS on-board data must be generated with a level of accuracy commensurate with the SIL 4 safety integrity level, which is required for the most critical functions of ETCS. This implies very rigorous processes for determining what data is required, verifying its correctness and for validating the behaviour of the system with it installed.

27.8 New build and retrofit

All of the above applies to any vehicle to be fitted with ETCS, but the cost and timescale for the fitment process varies a great deal, depending on whether ETCS is being fitted to a new train during its construction or is being retrofitted to an existing train.

ETCS is now sufficiently well established that all the major manufacturers of trains for the European market have already designed their rolling stock platforms to include ETCS, with space allocated for the equipment and connections. When the customer for a new fleet of trains specifies that ETCS is required, it can easily be incorporated into the design alongside all the other features that are needed. The validation and testing of the ETCS installation then takes place in parallel with testing of the other train systems.

Retrofitting of ETCS to an existing fleet of trains is a much more problematic process. Finding space for the equipment is usually difficult and often requires some existing train systems to be relocated. Running the cables to connect the new systems can be very intrusive. Existing power supplies may be inadequate in terms of capacity or quality. Depending on the age of the trains, there may be undocumented variations between vehicles of a nominally uniform type. Another issue is that the trains need to be taken out of service for the installation work to be undertaken. To manage the risk in these projects, it is important to undertake a 'first of class' installation for each vehicle type as early as possible in the programme, to identify all the generic problems and solutions before starting the main fleet fitment.

References

SUBSET-036: FFFIS for Eurobalise

SUBSET-040: Dimensioning and Engineering rules

Chapter 28: Testing and Certification of Trackside and On-board Subsystems

28.1 Testing of trackside and on-board installations

The complexity of ETCS means that extensive testing is required to ensure the system will be safe and reliable, and to guarantee interoperability between trackside and on-board subsystems from different suppliers. Testing with real trains on real infrastructure is essential for final validation, but unless a new railway is being constructed the time available for testing on an operational railway is severely limited.

It is therefore desirable to undertake as much testing as possible in laboratory conditions, and suppliers, infrastructure managers and independent laboratories have all invested in facilities designed for this purpose. Some of the testing can be performed separately for on-board and trackside in an extension of the environment used for testing the ETCS constituents (see chapter 24), by configuring the hardware under test with the parameters and interfaces for the specific subsystem application. Nevertheless, experience with practical projects has shown that it is also essential to test compatibility in a laboratory that incorporates both on-board and trackside subsystems.

The typical architecture for such a laboratory will allow a mix of real hardware and software simulations of ETCS subsystems that can be configured to simulate the operation of a specific route with a variety of trains. This setup is particularly useful for ETCS System Compatibility (ESC) testing, allowing many different on-board subsystems to be tested for compatibility with the specific ESC type. In laboratories equipped with multiple RBCs it is also possible to test the RBC-RBC handover functionality.

Where equipment from different suppliers is being connected in a laboratory there will understandably be concerns about commercial confidentiality, so it may be necessary to place equipment in different rooms, or to design the interfaces so that some of the components under test can be located remotely, for instance on the supplier's own premises.

Although a test laboratory can be configured to simulate the ETCS track-train interface, testing of the train on real ETCS-fitted track remains essential, and a dedicated test track is very useful for this purpose. Test tracks are already used extensively for testing new rolling stock more generally, and many of these have now had ETCS trackside systems installed to allow them to be used for ETCS testing of both new ETCS trains and existing trains retrofitted with ETCS. The configuration of the test track for the relevant ESC types should allow all the on-board subsystem testing to take place on the test track. An international train may need to be tested on more than one test track to complete all the applicable ESC/RSC tests.

Careful planning of the test campaign is very important, not only to identify which environment is appropriate for each test, but also to manage the logistics of access to the laboratory and test track facilities, which may be in high demand for testing of multiple projects.

28.2 Subsystem certification and authorisation

The remainder of this chapter describes the mandatory certification and authorisation process applicable to both trackside and on-board subsystems that are to be placed into service (operated) within the European Union. Other countries and railway administrations may apply similar processes as a matter of good practice, but they are not obliged to follow the EU processes described.

The TSI CCS defines a process for the certification of the on-board and trackside subsystems, equivalent to that for certification of interoperability constituents (described in chapter 24).

Whereas the manufacturer is responsible for constituent certification, subsystem certification is the responsibility of the contracting entity (who is responsible for the subsystem design and implementation and putting it into operation). The contracting entity for trackside subsystems is the relevant infrastructure manager, and the owner or operator of the trains is the contracting entity for the on-board subsystems (this is usually a railway undertaking, but it could be a train leasing company). The contracting entity must engage a Notified Body to perform an EC verification of the subsystem, and when this is complete the contracting entity issues an EC Declaration of Verification for the subsystem, supported by a Technical File and the EC Verification Certificate from the Notified Body. This is submitted to the European Railway Agency (ERA) or the National Safety Authority to obtain the authorisation for entry into service.

The purpose of the subsystem EC verification process is to verify that:

- all mandatory functions applicable to the subsystem (on-board or trackside) have been implemented.
- all optional functions that are required for the subsystem's intended specific application have been implemented.
- any additional functions implemented in the subsystem are not in conflict with the implemented mandatory and optional functions (i.e., they do not adversely affect the interoperability of the subsystem).
- the interoperability constituents of the subsystem have a valid EC Declaration of Conformity.
- the subsystem has all the EC declarations that are required by any other applicable European directives.
- for an on-board subsystem, all the ETCS and radio system compatibility (ESC/RSC) checks applicable to the intended area of operation (as declared in an ESC/RSC statement from the contracting entity) have been satisfactorily completed.

The work of the Notified Body includes:

- Checking documents and test results to verify that all the conformity assessment requirements specified in the TSI have been completed satisfactorily.
- For interoperability constituents within the subsystem, checking that their EC Declarations of Conformity are valid, that any conditions or limits of use are complied with, and the constituents have been correctly integrated into the subsystem. To do this, the Notified Body will require direct and unlimited access to the constituent's technical file.
- Verifying that the coverage of the ESC/RSC checks (whether performed at subsystem or constituent level) is adequate for the intended application.
- Assembling a technical file containing (or referencing) all the evidence of the subsystem's conformity with the TSI CCS, including conformity certificates for interoperability constituents used in the subsystem.
- If successfully verified, issuing an EC verification certificate for the subsystem.

The type and extent of the verification and checking activities performed by a Notified Body depends on 'certification modules'. All the certification modules are defined in the European Commission Decision 2010/713, and the TSI CCS specifies which ones are to be used and how they should be applied for the CCS subsystems.

There are four modules available for CCS subsystems, and the choice of which one(s) are to be applied for a given certification is subject to discussion and agreement between the contracting entity and its Notified Body. The choice depends upon the extent to which the

Notified Body can take account of the quality management system of the contracting entity and its supplier(s). The modules applicable to CCS subsystems are:

- Module SG (unit Verification)
- Module SB (type examination) + SF (product verification)
- Module SB (type examination) + SD (production quality management system)
- Module SH1 (full quality management system with design examination)

The options involving type examination or design examination are most applicable to the on-board subsystem, where the same subsystem is being installed in the same configuration on all the trains in a fleet. By contrast, each trackside subsystem is a one-off application for a specific area of railway, for which the concept of type examination is not really meaningful.

When the Notified Body has successfully completed its work and issued an EC verification certificate (which may, incidentally, contain caveats and constraints relating to the use of the subsystem), the contracting entity submits an EC Declaration of Verification to obtain authorisation for entry into service.

For a trackside subsystem, the Declaration of Verification is submitted to the National Safety Authority of the country concerned, but it must be preceded by an 'ERTMS Trackside Approval' from the ERA, which is a check on the interoperability of the trackside subsystem.

For on-board subsystems fitted to trains that are to operate within the EU, authorisation is via the ERA 'One Stop Shop' process. For trains that will operate in more than one EU country, the authorisation must be sought from the ERA itself. For trains that will only operate in one EU country, the contracting entity can choose whether the authorisation will be sought from the ERA or from the relevant National Safety Authority.

Chapter 29: Operator Training and System Support

29.1 Driver Training

For a train driver who is familiar with lineside signals and simple national train protection systems, driving a train fitted with ETCS represents a significant change. This is true even when driving a fitted train in non-ETCS areas, because the driver has to interact with the ETCS DMI to enable the train to start running in level 0 or level NTC.

This means that driver training is a significant element to be considered in any migration plan from conventional national signalling to ETCS. Driver training typically comprises three stages:

1. Driving an ETCS-fitted train on existing infrastructure in level 0 or NTC.
2. Driving a train on ETCS infrastructure (generic skills, not route-specific).
3. Route learning for a specific ETCS-fitted route.

For a new train delivered by the manufacturer with ETCS already fitted, the first stage will usually be integrated into the general training programme for that class of rolling stock.

The first stage is important in the early stages of migration, because training sufficient drivers to this standard is a precondition for the continuing operation of the trains after they have been fitted with ETCS, even though they are not yet running on ETCS infrastructure. Furthermore, it may be that the fitted trains operate much more widely over the rail network than just the ETCS-fitted routes and need to be driven by drivers who will never work on those fitted routes. For example, on the first ETCS project in Great Britain, 500 drivers required the first stage of training because they would have to drive fitted trains on unfitted routes, but only 45 required all three stages (because they were going to drive on the ETCS-fitted route).

The topics covered in the first stage of training provide an introduction to ETCS and what the driver needs to know in order to drive the train in level 0 or NTC in areas not fitted with ETCS. This includes:

- Overview of ETCS
- Trackside equipment
- New equipment fitted to the train
- How ETCS works
- ETCS levels
- How the DMI menus are accessed
- How to enter data at start of mission
- Modes and transitions
- Faults and failures
- Driving the train in level 0/NTC

The second stage extends the training to include driving the train in ETCS-fitted areas. For an area equipped with ETCS with level 2 cab signalling (no lineside signals) this would include:

- Start of mission in level 2
- Trackside signs in cab signalling areas and at transitions
- Operational modes associated with level 2
- Level 2 shunt areas
- Speed restrictions and how they are applied/enforced in level 2
- Faults and failures
- Written orders
- Driving a train in level 2
- Operation with ATO (where fitted)

The final (third) stage of training is to learn how ETCS has been applied on a specific route. The training may include some or all of the following, depending upon the driver's existing knowledge of the route:

- Track layout and location of stations and junctions
- Nomenclature of running lines and locations
- Location and spacing of signals (where provided) and block marker boards
- Permissible speeds (enforced by ETCS)
- Locations and layout of transitions into and out of ETCS
- Locations where specific modes may be applicable (OS, SH, AT)

The traditional approach to training a driver was for them to work alongside an experienced driver and, when ready to do so, take control of the train under supervision. Railway undertakings now use cab simulators, enabling most of the training to take place in a safe learning environment. However, both real trains and cab simulators are expensive resources to use, so the most efficient approach to ETCS driver training is to use more basic tools as far as possible, so that drivers already have a good understanding of the principles before they get 'hands-on' experience on a simulator or a real train. A typical training programme will therefore include elements such as:

- Traditional classroom presentations and handouts
- Interactive tabletop exercises
- E-learning modules on a computer
- Desktop simulation of ETCS DMI
- Full cab simulator with ETCS functionality
- Driving a real train under supervision

It is important to ensure that the training material and the simulators are kept up to date with changes to the software versions and application details of the real ETCS systems as these evolve during the application design and validation process.

For ETCS programmes and projects involving more than one type of train, there will need to be variants of the training to account for the different types. For larger fleets, a dedicated cab simulator for each type will be appropriate, but for small fleets such as track maintenance machines, a generic simulator would be used, followed by longer sessions driving the real train.

Where ATO is implemented with ETCS, the maintenance of driver competence for non-ATO driving, specific human factors issues associated with transitioning into and out of ATO areas, and the potential for loss of concentration on residual driver tasks in ATO areas, all need to be taken into account.

The timely provision of the training is critical, to ensure that sufficient drivers are available when ETCS-fitted trains and infrastructure are put into service. This is made more challenging by:

- The need to ensure that sufficient drivers are still available to operate the train service during the training period.
- The need to avoid a long period of time between the completion of training and using the new skills on a regular basis.

Figure 29-1: Driver training in a cab simulator (photo: GB Railfreight)

29.2 Control centre operator training

The ETCS training requirement for the staff in control centres is less extensive. The main operator (signaller or dispatcher) interaction with ETCS is via the traffic control system and the interlocking, in much the same way as with conventional lineside signalling. There may be some additional information provided on the workstation screens, and perhaps an additional terminal for interfacing with the RBC, for functions such as applying and removing temporary speed restrictions. A typical syllabus for control centre staff would include:

- Overview of ETCS
- How ETCS works
- Different ETCS levels
- Modes and transitions
- Changes to signalling principles
- Applying and removing temporary speed restrictions and emergency stop orders
- Faults and failures
- Written orders

Training for modern traffic control systems typically uses a training simulator, with a workstation replicating the real user interface, connected to a simulation of the interlocking, trackside equipment and train movements, with a facility for the trainer to apply and remove faults and deviations from the timetable. The whole simulation system usually also includes a simulation of the voice radio and telephone facilities, to allow the trainee and trainer to role-play a signaller to train driver conversation.

For an ETCS project, the training simulator needs to be upgraded so that the additional ETCS related controls and indications are realistically replicated. This will include simulation of an ETCS failure on a train, so that the training can include the use of voice communications and written orders.

29.3 Maintainer training – trackside

Training will also be required for the infrastructure manager's maintenance staff. There are two elements to this, for the trackside equipment and for the control centre equipment.

One of the aims of ETCS is to reduce the amount of equipment on or near the track, thereby reducing the trackside maintenance activity. The only new components on the track for ETCS are the Eurobalises and associated LEUs (the latter are principally required in level 1 areas only). While these are intrinsically reliable, there are many of them and they operate in a harsh environment, so occasional failures are to be expected. The maintainer training will therefore need to include:

- Interpreting failure information and diagnostic reports in order to identify a faulty balise
- Configuring a replacement balise
- Installing the balise correctly on the track
- Required system testing after any unit replacement or software/data change

Most infrastructure managers have training schools that are equipped with examples of all the trackside equipment in use on their networks, and an example of a balise installation with all the associated diagnostic, configuration and installation tools is a logical addition to this.

The control centre's ETCS equipment comprises the RBC, the associated hardware that connects it with the interlocking, traffic control system and RMR telecommunications systems, and any additional computer systems provided by the supplier for diagnostic and support purposes. The skill set required to maintain this equipment will be similar to that for modern electronic interlocking systems. The training will need to cover:

- Analysis of logged information to investigate failures and incidents
- Diagnosis of faults to identify the unit to be replaced
- Configuring and installation of replacement components
- Installation of software upgrades and data configuration changes
- Entry of data such as temporary speed restrictions (unless this is the responsibility of the operating department)
- EuroRadio key management (offline or online)
- Required system testing after any unit replacement or software/data change

A suitable training environment would be an RBC and associated interlocking equipment, with a system to simulate the movements of ETCS trains. This is very similar to a test laboratory used for RBC testing, and it may therefore be possible to use the same installation for both purposes.

29.4 Maintainer training – on-board

ETCS places much more of the signalling functionality on the trains (compared with conventional lineside signalling), and this has implications for the training of rolling stock maintenance depot staff. While the detailed signalling logic within the system is not relevant to them, they do need to understand enough of the system to diagnose problems and replace faulty modules. Their training will need to cover:

- Analysis of logged information to investigate failures and incidents
- Diagnosis of faults to identify the unit to be replaced
- Configuring and installation of replacement components
- Installation of software upgrades and data configuration changes
- Entry of data that needs updating for changes to the vehicle, e.g. wheel diameter
- EuroRadio key management (offline or online)
- Required system testing after any unit replacement or data/software change

29.5 System support

The complexity of a signalling system incorporating ETCS, and the interaction between the trackside and on-board elements, is such that there are bound to be system failures and unexpected behaviours that cannot be foreseen in the training of the first line operators and maintainers. It is therefore vital that a support organisation is established so that problems can be escalated and the root causes of issues and anomalies resolved.

During a deployment project this support may be provided from within the project delivery organisation, but it needs to be maintained on a long-term basis after the project finishes, throughout the operational lifetime of the systems. The infrastructure manager may take the lead in providing a permanent 'home' for the support organisation, but the continued involvement of other stakeholders and sources of expertise such as the railway undertakings and suppliers is essential, and the appropriate commercial and contractual arrangements will need to be in place to ensure this continues. Ideally, this will include ongoing access to the system test laboratories that were used in the initial development, to allow simulation of reported problems in an offline environment.

A key tool for the support organisation will be a cross-stakeholder Defect Recording and Corrective Action System (DRACAS). This is a set of processes and software tools for collaboration and information sharing in order to manage system failures, faults and defects. The DRACAS will typically include:

- Notification of an event from an operator and/or maintainer
- Recording factual evidence, status information and qualifying descriptions of the events
- Identifying and informing other affected or implicated organisations
- Investigation by one or more of the organisations
- Identification and implementation of containment and corrective actions
- Closure of investigation and measurement of corrective action

Such a system requires communication and collaboration between parties such that the root cause of issues and anomalies can be determined. A key element of a DRACAS system is ensuring that all parties have a common understanding of the system configuration. Where issues or anomalies are discovered, it is possible that, from an overall railway system perspective, the most appropriate solution will involve action by parties that were not directly associated with the original problem. That means that appropriate contractual and payment arrangements will also need to be in place to facilitate corrective action without undue delays.

It is quite likely that the organisations involved have their own in-house DRACAS systems which they use to manage all their assets or products, and so the ETCS DRACAS will need to be configured to exchange information with these other systems. Agreement on consistent terminology for describing and classifying issues and their consequences is very important.

Another element of system support will be the management of change through the lifetime of the ETCS trackside and on-board systems. Updates to system data and configuration, affecting both software and hardware, may be required for several reasons:

- Design corrections and enhancements required as a result of in-service experience (including experience gained from other ETCS applications using similar equipment)
- For the trackside ETCS, changes to track layout and stations
- For the on-board ETCS, changes to rolling stock characteristics
- Changes required as a result of ETCS standards updates, i.e. a new system version
- Changes required to address obsolescence of hardware or software

Any change to one ETCS component needs to be managed in such a way that it remains compatible with the systems it is interfaced to, and track to train interoperability is maintained.

One specific issue that will be a major consideration for European railways using ETCS level 2 in the 2030s will be the need to migrate the railway mobile radio system (RMR) from GSM-R to FRMCS, as described in chapter 18.

List of Abbreviations

3GPP	3rd Generation Partnership Project
ACC RBC	Accepting Radio Block Centre
AD	Automatic Driving (ETCS mode)
AEIF	European Association for Railway Interoperability (Association Européenne pour l'Interopérabilité Ferroviaire)
ARS	Automatic Route Setting
AsBo	Assessment Body
ATO	Automatic Train Operation
ATO-OB	ATO on-board
ATO-TS	ATO trackside
ATP	Automatic Train Protection
BIM	Building Information Modelling
BTM	Balise Transmission Module
C-DAS	Connected Driver Advisory System
CAN	Controller Area Network
CBTC	Communications-Based Train Control
CCS	Control-Command and Signalling
CENELEC	European Committee for Electrotechnical Standardization (Comité Européen de Normalisation Électrotechnique)
CER	Community of European Railway and Infrastructure Companies
CS	Circuit Switched (GSM-R mode of operation)
CSM	Common Safety Method
CTC	Centralised Traffic Control
CTCS	Chinese Train Control System
CW	Continuous Wave
DAS	Driver Advisory System
DES	Data Encryption Standard
DMI	Driver Machine Interface
DRACAS	Defect Recording And Corrective Action System
DTO	Driverless Train Operation
EAL	ERTMS Accredited Laboratory
EBD	Emergency Brake Deceleration curve
EBI	Emergency Brake Intervention
ECN	Ethernet Consist Network
EEIG	European Economic Interest Grouping
EIM	European Railway Infrastructure Managers

EoA	End of Authority
ERA	European Union Agency for Railways
ERRI	European Rail Research Institute
ERTMS	European Rail Traffic Management System
ESC	ETCS System Compatibility
ETCS	European Train Control System
EU	European Union
EUG	ERTMS Users Group
FFFIS	Form Fit Functional Interface Specification
FIS	Functional Interface Specification
FRMCS	Future Rail Mobile Communication System
FRS	Functional Requirements Specification
FS	Full Supervision (ETCS mode)
GNSS	Global Navigation Satellite System
GoA2	Grade of Automation 2
GoA3	Grade of Automation 3
GoA4	Grade of Automation 4
GPRS	General Packet Radio Service
GPS	Global Positioning System
GSM	Global System for Mobile Communication
GUI	Guidance (braking curve)
HDLC	High-level Data Link Control
HOV RBC	Handing Over Radio Block Centre
IC	Interoperability Constituent
IM	Infrastructure Manager
IPR	Intellectual Property Right
IRSE	Institution of Railway Signal Engineering
IS	Isolation (ETCS mode)
ISO	International Organization for Standardization
JRU	Juridical Recording Unit
KER	KVB, EBICab and RSDD train protection systems
km/h	kilometres per hour
KMAC	Key for Message Authentication Code (EuroRadio)
KMC	Key Management Centre
KSMAC	Session Key for Message Authentication Code (EuroRadio)
KVB	Contrôle de Vitesse par Balises (train protection system used in France)
L1LS	Level 1 Limited Supervision

LEU	Lineside Electronic Unit
LIDAR	Light Detection And Ranging
LoA	Limit of Authority
LRBG	Last Referenced Balise Group
LS	Limited Supervision (ETCS mode)
m	metre(s)
MA	Movement Authority
MAC	Message Authentication Code
MNO	Mobile Network Operator
MRDT	Most Relevant Displayed Target
MRSP	Most Restrictive Speed Profile
MVB	Multifunction Vehicle Bus
NID	ETCS Identity Number
NL	Non Leading (ETCS mode)
NoBo	Notified Body
NP	No Power (ETCS mode)
NTC	National Train Control
OS	On Sight (ETCS mode)
OSI	Open Systems Interconnection
PKI	Public Key Infrastructure
PS	Packet Switched (GSM-R mode of operation)
PS	Passive Shunting (ETCS mode)
PT	Post Trip (ETCS mode)
PTC	Positive Train Control
QoS	Quality of Service
RaSTA	Rail Safe Transport Application
RBC	Radio Block Centre
RFU	Recommendations For Use
RMR	Railway Mobile Radio
RRI	Route Related Information
RSC	Radio System Compatibility
RSDD	Ripetizione Segnali Discontinua Digitale (train protection system used in Italy)
RSSI	Received Signal Strength Indication
RU	Railway Undertaking
RV	Reversing (ETCS mode)
S-DAS	Standalone Driver Advisory System
SB	Stand By (ERTCS mode)

SBD	Service Brake Deceleration
SBI	Service Brake Intervention
SF	System Failure (ETCS mode)
SH	Shunting (ETCS mode)
SIL	Safety Integrity Level
SL	Sleeping (ETCS mode)
SM	Supervised Manoeuvre (ETCS mode)
SN	National System (ETCS mode)
SoM	Start of Mission
SPAD	Signal Passed At Danger
SR	Staff Responsible (ETCS mode)
SRS	System Requirements Specification
SSP	Static Speed Profile
STM	Specific Transmission Module
STO	Semi-automated Train Operation
SvL	Supervised Location
TAF	Track Ahead Free
TCP/IP	Transmission Control Protocol/Internet Protocol
TLS	Transport Layer Security
TMS	Traffic Management System
TR	Trip (ETCS mode)
TSI	Technical Specification for Interoperability
UIC	International Union of Railways (Union Internationale des Chemins de fer)
UML	Unified Modelling Language
UN	Unfitted (ETCS mode)
UTO	Unattended Train Operation
VBC	Virtual Balise Cover
VBR	Virtual Balise Reader
VoLTE	Voice over Long-Term Evolution
WSP	Wheel Slide Protection

Index

Advertisers Index